The Venice Arsenal
Between History, Heritage, and Re-use

This book reviews four decades of debate about restoring an industrial heritage site of inestimable value – the Venice Arsenal. Focusing on the challenges of economic, financial and institutional feasibility, it reveals how failing to address these aspects has undermined potential solutions from both technicians and heritage professionals.

With a deep connection to the city over centuries, the Arsenal was the very basis of *La Serenissima*'s sea power, enabling its economic expansion. Later, it maintained a vital military function through shipbuilding until World War II. But the slow process of abandonment of the traditional site's uses and spaces continues to pose questions regarding its preservation and re-use. Drawing on original research from urban planners, architects and historians, the book provides a critical investigation into the organizational and managerial challenges of this unique site, and crucially, why so little has been achieved compared with potential opportunities.

Featuring numerous color photographs and exploring the particular challenges of restoration and re-use facing the Venice Arsenal, this insightful evaluation of the history of this site provides a uniquely informative case for the discipline of industrial heritage.

Luca Zan is Professor of Arts Management at the University of Bologna, Italy, and Central Academy of Fine Arts, Beijing; he has been adjunct faculty at Heinz College at Carnegie Mellon University for several years.

Routledge Research in the Creative and Cultural Industries
Series Editor: Ruth Rentschler

This series brings together book-length original research in cultural and creative industries from a range of perspectives. Charting developments in contemporary cultural and creative industries thinking around the world, the series aims to shape the research agenda to reflect the expanding significance of the creative sector in a globalised world.

Managing Cultural Festivals
Tradition and Innovation in Europe
Edited by Elisa Salvador and Jesper Strandgaard Pedersen

Music as Labour
Inequalities and Activism in the Past and Present
Edited by Dagmar Abfalter and Rosa Reitsamer

Risk in the Film Business
Known Unknowns
Michael Franklin

Orchestra Management
Models and Repertoires for the Symphony Orchestra
Arne Herman

The Venice Arsenal
Between History, Heritage, and Re-use
Edited by Luca Zan

For more information about this series, please visit: www.routledge.com/ Routledge-Research-in-the-Creative-and-Cultural-Industries/book-series/ RRCCI

The Venice Arsenal

Between History, Heritage, and Re-use

Edited by Luca Zan

Routledge
Taylor & Francis Group
LONDON AND NEW YORK

First published 2022
by Routledge
4 Park Square, Milton Park, Abingdon, Oxon OX14 4RN

and by Routledge
605 Third Avenue, New York, NY 10158

Routledge is an imprint of the Taylor & Francis Group, an informa business

British Library Cataloguing-in-Publication Data
A catalogue record for this book is available from the British Library

Library of Congress Cataloging-in-Publication Data
Names: Zan, Luca, editor.
Title: The Venice Arsenal : between history, heritage, and re-use / edited by Luca Zan.
Description: Abingdon, Oxon ; New York, NY : Routledge, 2023. | Series: Routledge research in the creative and cultural industries | Includes bibliographical references and index.
Identifiers: LCCN 2022011218 (print) | LCCN 2022011219 (ebook) | ISBN 9781032059617 (hardback) | ISBN 9781032059624 (paperback) | ISBN 9781003200055 (ebook)
Subjects: LCSH: Arsenale di Venezia—History. | Arsenale di Venezia—Management—History. | Urban renewal—Italy—Venice. | Shipyards—Italy—Venice—History.
Classification: LCC VM299.7.I8 V46 2023 (print) | LCC VM299.7.I8 (ebook) | DDC 623.8/30945311—dc23/eng/20220315
LC record available at https://lccn.loc.gov/2022011218
LC ebook record available at https://lccn.loc.gov/2022011219

ISBN: 978-1-032-05961-7 (hbk)
ISBN: 978-1-032-05962-4 (pbk)
ISBN: 978-1-003-20005-5 (ebk)

DOI: 10.4324/9781003200055

Typeset in Times New Roman
by Apex CoVantage, LLC

Contents

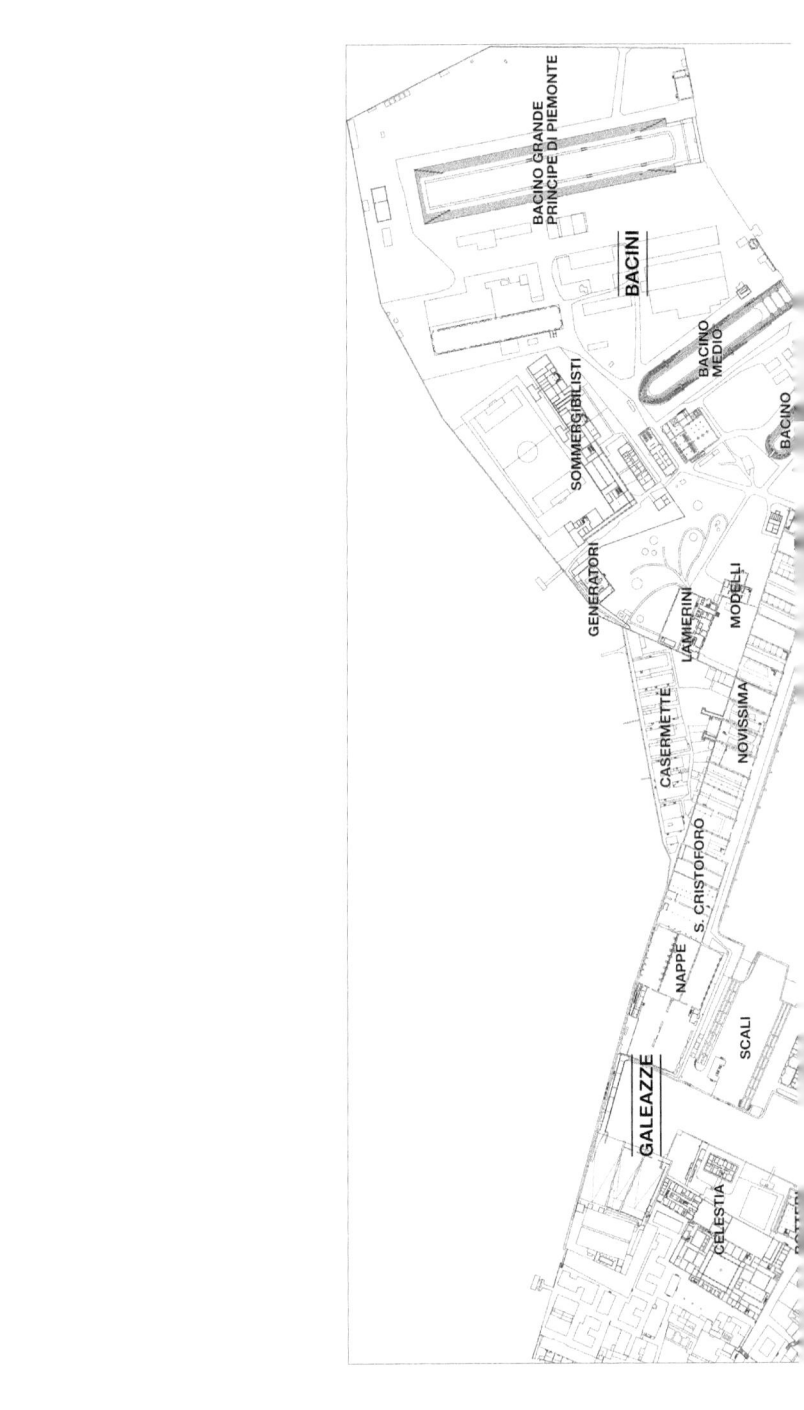

BACINO GRANDE
PRINCIPE DI PIEMONTE

BACINI

BACINO
MEDIO

BACINO

SOMMERGIBILISTI

GENERATORI

LAMERINI

MODELLI

CASERMETTE

NOVISSIMA

S. CRISTOFORO

NAPPE

SCALI

GALEAZZE

CELESTIA

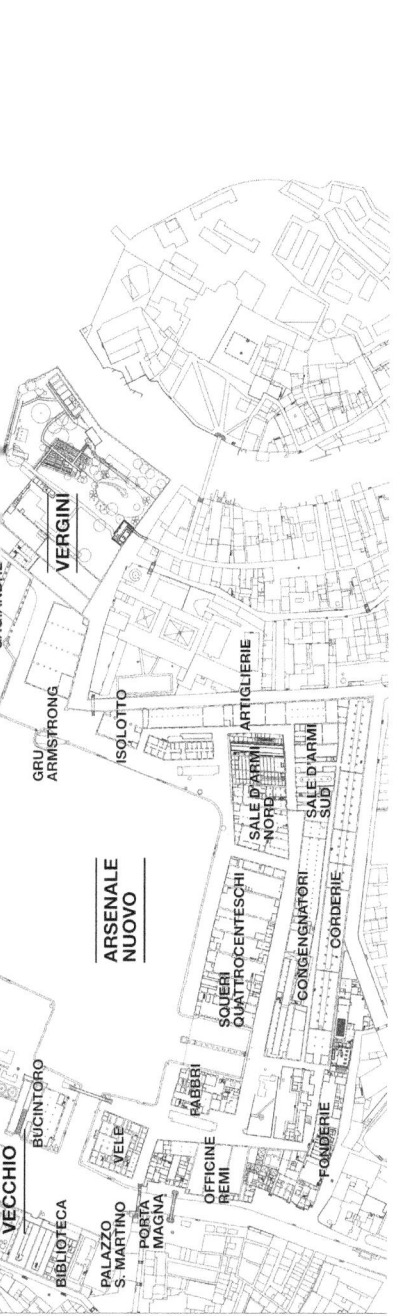

Figure P.1 Main buildings and areas inside the Arsenal

Acronyms

ACTV, Azienda Veneziana della Mobilità, public transport provider for the urban area of Venice

ARAC, Associazione per la Restituzione dell'Arsenale alla Città

COMAR, Costruzioni Mose Arsenale Spa, commercial arm of CVN

CNOMV, Costruzioni Navali e Officine Meccaniche di Venezia, private company Venice Shipbuilding and Machine Workshop

CNR, Consiglio Nazionale delle Ricerche, National Research Council

CSA, Centro Studi Arsenale

CVN, Consorzio Venezia Nuova, New Venice Consortium, Concession of MIT responsible for MOSE

FIO, Fondo per gli Investimenti e l'Occupazione, Fund for Investment and Employment

HBB, Harbour Brain Building

ISMAR, Istituto di Scienze Marine, Institute of Marine Science

ISMM, Istituto di Studi Militari Marittimi, Institute for Maritime and Military Studies

IUAV, Istituto Universitario di Architettura di Venezia, University Institute of Architecture, Venice

MAV, Magistrato alle Acque di Venezia, Venice Water Authority

MiBACT, Ministero per i Beni e le Attività Culturali e Turismo, ministry with responsibility for cultural property, heritage and the historic environment

MIT, Ministero delle infrastrutture e dei trasporti, Ministry of Infrastructure and Transport

MOSE, MOdulo Sperimentale Elettromeccanico, Venice Lagoon Floodgate System

PPAN, Piano Particolareggiato Area Nord, Detailed Plan – North Area

PRUSST, Programma di Riqualificazione Urbana e Sviluppo Sostenibile del Territorio, Program of Urban Regeneration and Sustainable Development in the Territory

SAV, Società Arsenale di Venezia, limited company owned jointly by the state (51%) and the Town Hall (49%).

Short glossary

Agenzia del Demanio: State Property Agency, government department with ownership and oversight of all state-owned property

Genio Difesa: Ministry of Defense Engineers

Forum Arsenale: Arsenal forum, a grassroots association (or citizens) to support the Arsenal

Soprintendenza   (Soprintendenza Archeologia, Belle arti e Paesaggio di Venezia e Laguna): Superintendency, local branch of MiBACT (Office for the Protection of Archaeological, Fine Arts and Landscape of Venice and the Lagoon) A.

Tesa nf *tese* pl: shed or roofed area, open-plan and either wholly or partially open at the side(s). They are numerous throughout the Arsenal, located variously on docks, at the waterside and elsewhere and serve numerous functions, for example, shipbuilding, maintenance/repair and storage. There is no single-word equivalent in English, so we have retained the Italian term throughout.

Vela, Ve.La Spa.,   commercial events and services subsidiary of ACTV

Contributors

Authors

Ettore Cammarata is an engineer and the former Director of the *Agenzia del Demanio* for Veneto, Friuli Venezia Giulia and Sicily and has worked on the use and valorization of public real estate assets. He is currently a freelance expert in real estate valuation.

Roberto D'Agostino is an urban planner in several Italian cities and European and non-European countries. He was Councilor of the Municipality of Venice between 1993 and 2005 and President of the SVA from 2006 to 2014.

Paolo Ferri is associate professor in accounting and GIOCA director (Graduate degree in Innovation and Organization of Culture and the Arts) at the University of Bologna. His research interests deal with the study of innovation in accountability systems among complex and professional organizations, mainly among cultural organizations in Italy and abroad.

Valentina Gambelli is a professional architect. She teaches and does research at IUAV University, focusing on the Venice Arsenal.

Pegram Harrison, University of Oxford, is senior fellow in entrepreneurship and director of undergraduate studies in management at the university's Saïd Business School. He also teaches on the Oxford Cultural Leaders program and conducts research on the leadership of cultural and heritage organizations.

Franco Mancuso is an architect. He has taught urban planning at IUAV, Venice, and has held seminars and conferences in European and non-European universities. He is active in research and design on urban recovery and redevelopment.

Claudio Menichelli is an architect. He has worked for MiBACT, at the Superintendence of Venice (1980–2011), where he was responsible for the Arsenal for many years. He also has taught restoration at IUAV, Venice (1997–2009).

Camillo Tonini has been conservator of historical and geographical collections of the Correr Museum in Venice and was responsible for the Doge's Palace, including the computerized cataloguing service and serving as director of the *Bulletin*.

Pasquale Ventrice is a historian of science and technology, former professor at IUAV, Venice, and President of the *Centro Studi Arsenale*. He is Director of the Venice section of AIPAI.

Luca Zan, University of Bologna, is active in international research on management and accounting history and management of arts and cultural organizations. He teaches Arts Management at GIOCA, University of Bologna, and Central Academy of Fine Arts, Beijing, and has been adjunct faculty for several years at Carnegie Mellon University, Pittsburgh.

Translator of chapters originally in Italian

Anthony Evans-Pughe studied French and Italian at Edinburgh University and is now a translator working in Oxford and London. He also trained as a stonemason and worked for many years in the south and west of England as a conservator of architectural stonework.

Introduction

Luca Zan

Aims of the book

The Venice Arsenal is an industrial heritage site of inestimable value, with a deep connection to the city over a long span of time. Known throughout Europe as *officina de'e maravegie* (the workshop of wonders), for centuries it was the very basis of *La Serenissima*'s sea power, which enabled its economic and commercial expansion. It was extremely important during the golden age of the Venice Republic, and despite many changes over the centuries, it maintained a vital military function, essentially continuing to support shipbuilding activities until World War II.

Which brings us to today. A slow process of abandonment of some of the spaces and military uses poses new questions regarding the recovery and re-use of the site. While these issues are not uncommon at similar sites, some problems are unique to the Arsenal. This is due in part to the cost of securing and re-using the complex in its entirety and the technical challenges posed by the size of individual buildings. Many projects have been undertaken during the postwar decades, alternating with periods of inattention and neglect, all in the context of the complex problems of such a unique city, and with the tensions and risk of being included on the list of endangered UNESCO sites.

This volume is placed within this field of work. Its contributions partly constitute work presented at two Ateneo Veneto conferences in 2017 and 2018 which is already available in Italian (Zan, 2018, 2019a), with some additions for an international audience.

The book seeks to operate on three levels. First, while it is based on documented sources, it seeks to present this popular topic for a wider audience than just specialists.

Second, given the "universal" value of the site (e.g., Fletcher and Spencer, 2005), it aims to bring international attention to the Arsenal (and to Venice) and to expand interest in the site beyond occasional stories in the local press.

DOI: 10.4324/9781003200055-1

From this point of view, it is surprising that in the recent tension between UNESCO (2016a, 2016b) and the City of Venice (Città di Venezia, 2017) about the possibility of listing Venice as an endangered site, there is no discussion at all of the Arsenal, except for a quick reference to the control room of MOSE. More in general, it is also disconcerting that years of discussion and planning have simply been forgotten. Thus, the book seeks to reconstruct a 40-year discussion and to reflect on the reasons for this continuous elision and lack of impact.

A third aim is to provide an organizational–managerial investigation. In fact, the question of "management", and the economic, financial and institutional feasibility of possible re-uses of the Arsenal, is one of the least studied aspects of this story. Lack of attention to this realm could easily condemn to failure all the efforts of technicians and cultural heritage professionals. There is thus a need to reconstruct the costs and funding of work carried out so far, as well as to reflect on the financial, organizational and institutional conditions for the future sustainability of any project to restore and re-use the Arsenal.

The Arsenal: between Venice's past and future

The historical value of the Arsenal, and the set of meanings historically associated with it, is immense: from the point of view of social history (Lane, 1934), of architectural history (Bellavitis, 1983; Concina, 1984), of economic history (Romano, 1954; Davis, 1991), but also of the less well-known field of "management history" (Zan, 2004a). Moreover, starting from the second half of the 19th century, the Arsenal underwent a deep process of industrial reconversion, which gave it interesting elements of advanced industrial production, a sort of second youth that lasted almost a century.

The *first part* of this volume focuses on these aspects, in a very selective way (indeed, reconstructing the long process of evolution of the Arsenal over 900 years would require more than a book on its own: for a preliminary chronology see Concina, 2004). The two contributions investigate two major discontinuities in the administrative history of Arsenal, which explore intangible aspects of this industrial heritage site in relation to specific problems of "valorization".

Chapter 1 reconstructs the significance of management practices at the Venice Arsenal at the turn of the 16th century and the contribution to the development of modern management knowledge. It shows the development of a sophisticated conversation about management *(discorso del maneggio)*, which challenges several assumptions of economic, business and accounting history.

Chapter 2 provides an overview of the development of the Arsenal's production between the 19th and 20th centuries, with particular regard to skills and the organization of labor following the technological shift from wood-to iron-based production.

However, the present volume is not a historical contribution per se. It also seeks to address the question of how to intervene today to preserve this industrial heritage site. In this regard, the reconstruction of the interventions carried out is an important element for future policies. Understanding the past – in this case, the less studied recent past: the past 30–40 years – in order to manage the future. This provides not a generic statement of common sense but a perspective that calls for important arguments. The *second part* of the volume focuses on these aspects.

Phenomena of deindustrialization (and, with slight variations, of demilitarization) are certainly not rare, and from this point of view the Arsenal is in good company. It is the same economic process of "creative destruction", as Schumpeter put it, that generates cases like this: decommissioned sites, supplanted by more competitive technologies, products or sometimes just locations. These sites are "naturally" destined to go out of business and history. Only an attitude that goes "against nature" can stop this decline. And the only way to keep part of the history is to sacrifice another part, to think of re-uses that have nothing to do with the original, or even previous, functions and structures.

This is not a recent problem. The meanings of the site have always been in motion. They are dynamic by nature: the continuous renovations from the Middle Ages onward have continually challenged previous functions, just as the transformations of the 19th and 20th centuries have profoundly modified skills, jobs, processes and structures. Basically, one could think of the Arsenal as the coexistence of dynamically composed meanings or, as suggested here, a stratigraphy of uses and meanings over time.

The problem of re-uses, of the process of transformation of meanings (Balzani, 2004, uses the neologism "re-semantization") is quite common in cultural heritage and specifically in industrial heritage. Just think of how many former ports or maritime sites have been transformed in the world in recent decades, giving rise to a "new life", a new life cycle of sites and remains that had already been condemned to death by the creative destruction typical of the processes of economic innovation, in the terminal phase of the previous life cycle. Indeed, the meanings of a site can change over time, even if its reinvention requires considerable resources and creativity. If this aspect is intriguing from a theoretical point of view – with new "creative" opportunities opposing the previous processes of destruction – there are important practical consequences for managing these creative processes. This is particularly the case with regard to governance and leadership.

There is, however, something specific to the Venice Arsenal, and identifying this specificity is essential for understanding the terms of future change.

The Arsenal has always been a large site – immense, in fact. These spatial characteristics are paired with an area of naval production that is among the oldest in the world, and which was still operational until relatively recently. And unlike many other sites, it has been preserved in its spatial grandeur.

Certainly, it is a spurious preservation, which has already transformed the site, as seen, for example, in the 19th- and 20th-century destruction of many the original artifacts. But the complex remains substantially intact – indeed, much more so than other ports or historical shipyards around the world (from Portsmouth to Rotterdam to Istanbul, just to name a few). While I am not particularly sympathetic to the concept of "cultural capital" by Throsby (2005), in this case, it may be worth considering the potential market value an area of this size could have, if only as an opportunity–cost concept (what we are giving up by not allowing the systematic destruction of what remains): as a rough calculation, 48 hectares at €4,000–5,000 per square meter is in the order of €2 billion!

In addition to the overall scale of the site, the number and size of the individual artifacts (e.g., the anomalous size of the *tese* themselves, significantly larger than other shipyards of the time) create problems for individual re-uses. Even if commercial exploitation were legally possible, the transformation of the site into buildings for residential use, as often happens in other contexts, would be impossible here. Paradoxically, it is the abundance of huge spaces and historically significant remains that represents a challenge rarely encountered elsewhere.

In this framework of the dynamic creation of meanings – which are never static and never definitively configured – the history of interventions at the site contributes to further idiosyncrasies, including the past 30–40 years: forms, functions and meanings are modified; new stakeholders emerge in parallel; and the perceived universe of possible actions changes. Take the case of the *Biennale* for example. No one in 1957, at the end of the Navy's intensive and exclusive use of the Arsenal, would have thought this a possible new use (and possible new stakeholder). However, from 1980 onward, the *Biennale* has been one of the main actors in the overall recovery process of the Arsenal and is today among the most important (and powerful) stakeholders.

Meanings, uses and stakeholders are therefore dynamic elements themselves (yet bearing in mind the potential abuses of stakeholder rhetoric, since their greed can be the most dangerous and disruptive element for the survival of a site in general, and perhaps particularly for the Arsenal). To understand the nature of these elements, it is vital to reconstruct recent interventions. On the one hand, much remains to be done, and urgently – as we know, the lack of use itself causes rapid degradation. But on the other, this need itself is only present because so much has been preserved over the centuries and in particular in the past three decades (in comparison with other similar sites, and even with all the technical and functional transformations). Understanding the logic of these problems – the partial degradation, the interventions of a partial but largely generous recovery, the unresolved issue of decisions about possible destinations, the plan of interventions with related costs and funding – all this requires a good understanding of what has been done so far. Only a detailed reconstruction of the processes underway can help us arrive at operational solutions within a reasonable time frame. And interestingly, the chapters in this second part of the book are written by people who were important actors during this time.

Chapter 3 provides a perspective on one of the central institutions in the recovery of the Arsenal in these decades: the Superintendency of Venice and the Lagoon. It outlines an overall design of the recovery plus its actual implementation, at least until 2013–2014, when there was an overall slowdown in interventions.

Chapter 4 yields an interesting insight into the processes of the search for efficiency in the management of State Property Agency *(Agenzia del Demanio)*, which in Venice has a special position due to the history of ownership and its recent developments.

Chapter 5 gives a privileged testimony from the perspective of one of the central institutions in a phase of the Arsenal's revival process, the *Società Arsenale di Venezia* (SAV) company. The tone is strongly critical, but the issues that the author emphasizes go to the heart of the debate and controversies surrounding the development of the Arsenal (and the events of SAV itself).

Chapter 6 investigates some of the main contradictions in the process of recovery up to now while looking at the process of protection and design of potential re-uses.

The problem of access, and of the rights of citizens and local communities, is a central part of recent developments in the discourse on cultural heritage at an international level. And here, too, the rights of the local community are affirmed, which involves alternative uses of the site while claiming free access and explicitly accusing other stakeholders of being obstacles to this new vision. For example, the part of the Arsenal under control of the *Biennale* is accessible only for six months a year with payment of a €20 fee and closed for the rest of the time. The *third part* of the book focuses on these aspects.

The topic of the Arsenal Museum is not new. On the contrary, it has been talked about for 40 years, even though in reality nothing has been done. The first discussions began in the 1970s (e.g., Chirivi, 1976) and developed over subsequent decades (Lombardi and Paternò, 1992, p. 1, referred to "twenty years of projects on the Arsenal"). What is surprising is the overall opacity of the debate around the various projects and the instability of the "museum issue" over time. Rather than discussion and controversy about different solutions, the various proposals essentially ignore each other in a sort of a game of plans. From an organizational (and policy) perspective, important processes that would allow learning from mistakes are not activated, such as selecting priorities and preferences in a more considered, transparent and stable way. As for the inconsistency over time, if the 2001 Master Plan did not include discussions about an Arsenal Museum that were developing in those years, in the recent versions of the Master Plan (Città di Venezia, 2014, 2015) every reference to the museum has disappeared, simply ignoring the rich debate we will discuss in this volume. Indeed, for the past ten years, the question of the museum seems to have been completely removed from the debate on Venice and on the Arsenal re-uses itself. Here then is the first reason, analytical we might say, for revisiting the question: the need to compare hypotheses and projects that are often totally separate; the identification of possible intrinsic weaknesses that led to its abandonment by possible opponents; the emergence, nonetheless, of a common feature of the whole discussion of those years (from 2000 to about 2008), which then suddenly disappeared: the idea that the museum should be accompanied by the total opening of the Arsenal to citizens.

And here lies the second reason, in the current situation of total oblivion, in which the city's institutions seem to have removed the problem of the Arsenal itself in recent years – not only has the museum disappeared, but so has the whole issue, including the Master Plans. In the absence of any willingness to find a solution to the overall recovery of the Arsenal by institutions involved, our aim is to re-propose the theme of the museum as the central element of a serious reconsideration of the Arsenal. It is not just a question of drawing on knowledge and contributions, important though that is: the reopening of the debate on the museum is a "lockpick", to use a Gramscian term, for reopening the debate on the Arsenal more generally – or rather, for opening the Arsenal and giving the public access to it.[1]

Chapter 7 introduces some interesting reflections about the collections that could be used in the hypothesis of opening the Arsenal Museum, in addition to what has already been outlined by various projects. This would make it possible to strengthen the narrative structure of the Arsenal, recovering historical materials already present but not yet (or no longer) presented by city museums currently.

Chapter 8 presents a review of the debate on the new Arsenal Museum since its origins, explicitly linking the problem of the museum to that of access, always in terms of openness to the public, as mentioned earlier.

Chapter 9 instead traces the events of the project of the National Museum of Naval History, prepared in 2007 by ISMM *(Istituto di Studi Militari Marittimi)*, in strong collaboration with other institutions, including the Superintendency, of which the chapter author was an official at the time, responsible for the Arsenal.

Chapter 10 provides an analysis from the perspective of management studies, focusing on the project *Arsenale e/è museo* (the Arsenal is/and the museum), as well as other two projects by CSA *(Centro Studi Arsenale)* and CNR (National Research Council) and ISMM, in search of the financial and management assumptions that characterized these projects, and the related issue of their problematic sustainability.

The final *fourth part* of the book highlights a possible research agenda for the development of a more robust approach to the issue of the recovery of the Arsenal.

Chapter 11 sums up some of the contradictions that need to be resolved to recenter attention on the Arsenal, with a particular focus on the overall weakness (if not total absence) of a managerial perspective in the whole debate about studying, protecting and presenting the historical meanings of the Arsenal as intangible heritage.

Chapter 12 proposes the preliminary results of a comparison between the Venice Arsenal and other European historical shipyards. The comparison with similar sites is a central theme if we want to learn from good practices in other contexts and enhance the recovery of the Venice Arsenal, without losing sight of the historical, structural and functional differences.

Chapter 13 provides a more general framework to guide a future comparison across historical shipbuilding sites within an interdisciplinary perspective, though not forgetting the management angle.

A short conclusion follows, with a "policy agenda", in parallel with the research agenda of part IV, that calls for reclaiming the huge amount of work done in these 40 years, with all possible corrections and integrations.

Note

1 An investigation into civil society associations that were active in the recent period in Venice would be worthwhile, particularly the proposals developed by the Forum Futuro Arsenale (2016). We are not entering the contents of these proposals in this book because the main objective is to investigate reasons and mechanisms of inaction by main Venetian institutions, more than following reactions to such a situation. By the way, as part of this inaction, the contents themselves of the Forum proposal have been largely ignored by the Municipality.

Part I

A reminder

On the importance of the Arsenal
in the history of administration

1   History of management and stratigraphy of organizing

Luca Zan

The Venice Arsenal: from industrial heritage to management history

This chapter looks at the Venice Arsenal as an economic organization engaged in state manufacturing and the history of its organizational processes. In particular, it focuses on the period between the Battle of Lepanto (October 1571) and the end of the 17th century.

When examining this kind of organization, we normally consider a variety of production factors and conditions: the buildings themselves, machinery and technologies, materials, workers with their skills and nature, and the ways in which all this is organized. Unfortunately, the whole picture tends to evolve quickly and profoundly, and configurations at specific points in time often vanish. People, machinery, final products and raw materials disappear. What remains, in some cases, is simply the walls and the buildings themselves. Writing management and organizational history is incredibly complex in this context, especially when dealing with an organization like the Arsenal, which was active for almost 900 years.

Sometimes, we are lucky and discover products from the past: this was the case with the discovery of a 13th-century galley in the Venice lagoon in 1996 (the *Boccalama* galley: see Chapter 9 and Figure 9.1). However, even machines disappear quickly: the huge spaces across the whole complex of the Arsenal have lost almost all their equipment, even from recent periods, let alone rope machinery from the 16th century.[1] If material artifacts in general disappear as quickly as people, how can we write a history of organizing in this context? Sometimes, ancient paintings can provide insights into tools, operations and tasks, but these are particularly difficult to interpret when they are centuries old and represented by a completely different division of labor (see the wonderful paintings available at the Correr Museum, some of which are presented in Figure 1.1).

DOI: 10.4324/9781003200055-3

And regardless, it is very difficult to infer management practices and organizational logics from material remains. If management is about "getting things done", looking at the history of management or the history of organizing means to look at the process of "how things get done". Alternatively, if we understand management as an issue of "addressing attention" (March, 1988), the question becomes about the history of how attention was addressed in the past (or in a specific period of time).

There is a very famous painting that is presented in almost every historical account of the Venice Arsenal: the work by Maffioletti in 1797 (Figure 1.2), portraying ships and materials stored in the huge area of the

Figure 1.1 Paintings of Venetian guilds at the Correr Museum, XVI-XVII century: rowers, caulkers, sewers (© Gallerie dell'Accademia, Venice, on concession by the Ministero della Cultura)

Figure 1.1 Continued

Arsenal, including rows of wood, cut timbers, rudders, oars, masts, ropes and artillery and so forth. The painting (and its intrinsic "order") cannot be understood without historicizing its genesis. It is not the product of a "natural" human impulse to conceive and represent this type of order. Instead, it can be seen as the result of organizational practices and at least two centuries of conversations about defining layout and productive spaces, as well as operation and labor control, that we can find in available sources in the archives. In this way, we can link texts and material remains of the old site.

Old and new sources: a stream of systematic reports on the management of the Arsenal (1581–1781)

Within the debate on Venice history during the golden age of *La Serenissima*, there is vast agreement about the emergence of a period of administrative reform in the last quarter of the 16th century. However, historians tend to disagree about the meanings of this transformation, with two positions summing up the opposing views of this debate. According to Lane (1934), the modern characteristics of current industry were at stake, with the invention of the assembly line, interchangeability of pieces and vertical integration. According to Concina (1984, p. 175), on the other hand, administrative skills were lacking. Curiously enough, neither of them has – and none of

Figure 1.2 View of the Arsenal by G.M. Maffioletti, 1797 (ownership by Museo Storico Navale, Marina Militare)

the historians writing about the management of the Arsenal have –any background in organization, management or accounting. Thus, there are huge risks that management and organizational issues will be trivialized.

With the specific lenses of management/organizational/accounting history, we undertook a systematic screening of all sources at the Venice State Archive dealing with the Arsenal in the period 1540–1594. We discovered a fundamental deliberation of the Senate in 1580 that obliged the people in charge of the Arsenal to account for production, specifically in regard to the establishment of a reserve of 100 galleys. In short, the main issues include the following:[2]

- The Battle of Lepanto in 1571 soon became a useless victory. Just six months later, the Ottomans launched a completely new fleet (Ari and Zan, 2021).
- In response, the Senate re-issued an old deliberation – which had not been taken seriously until then – to build a reserve of 100 light galleys plus 12 great galleys to use in the case of war.
- This was a challenging goal, hard to achieve. This is why the Senate imposed a form of reporting in relation to the Arsenal and the achievement of the 100 plus 12 goal.

What we found in the archive, more than the four documents referred to in the literature (two by Drachio and two by Tadini), was a stream of reports, starting in 1580 and ending in 1781. Most focused on reporting on the situation of the 100 plus 12 galleys but then reflected on general issues of organizing and managing the entity. In this regard, what is extraordinary in Venice is not the invention of management per se, but rather, that there is a reflection on operations and management, a *discorso del maneggio* as it was called at that time: a conversation about organizing. That soon gave rise to new notions, concepts and a vocabulary about managing knowledge.

On the basis of these documents, we can infer management processes: problems, priorities and focuses of attention.

Management discourse at the Venice Arsenal, 1580–1713

There are three kinds of documents that illustrate the development of a management discourse at the Venice Arsenal during this period: the contributions of two internal officers (Drachio and Tadini) and the stream of periodic reports (starting in 1581 and ending in 1781, although the last reports have a completely different meaning, so we only consider reports up to 1713. For complete archival references see Zan, 2021).

Baldissera Drachio, 1586

Reading this 1586 document is a thrilling experience for a management scholar: more than 400 years ago, someone was grappling with the same issues that we do.

Drachio addresses several topics:

- Wood – the design of procedures for the acquiring wood (selection, transport, aging) and accessing the storage are discussed.
- A call for the "common timber", timbers in common lengths – a standardization of the measurements of the galleys, to simplify production and refurbishment, rather than tailoring each ship individually.
- How to structure spaces – allocating specific areas for new galleys and those being refurbished and ordering wood close to these areas according to different uses, to make retrieval easy and avoid worker downtime as they wait for material.
- Workers – a detailed design of six teams is provided, each with 30 carpenters and their main tasks, each described in detail (it is rare to find such detail in documents before the 19th century).
- The organization of those in charge – there is a call for the establishment of a new position of a single superintendent who could reform the Arsenal, which is a problem Drachio poses dramatically. Today, we are used to thinking in terms of an organizational "pyramid". That was not the case then: every patrician family had its leaders and protégés.

On the whole, the document was quite harmful for Drachio personally: he was exiled from Venice for ten years. Rather than merely redefining a production process, he was questioning the whole underlying social system.

Bartolomeo Tadini, 1593 and 1594

The first line in Tadini's document epitomizes his focus and background as the chief accountant of the Arsenal: "The huge expenses of the Arsenal consist of workers and materials".

- On the one hand, Tadini begins by talking about material and warehouses. He then moves on to discuss theft. In the guild tradition, waste and scraps were the property of the craftsmen. Tadini questions this and seeks to define the issue differently.
- On the other hand, Tadini proposes a series of suggestions to control the workforce by both checking attendance and providing incentives. In addition, since they were running behind in building the 100-galley reserve, Tadini proposed creating a temporary organization to build the

remaining 48 galleys in two years (six teams, asked to build two ships per semester for two years).

The argument is in itself fascinating: the reform was to be organized "according to a principle of order, emulation, rewards and punishments", echoing language we still use today in Italy to define the reward system. Finally, Tadini summarizes his proposal in terms of budget, suggesting that it will not only achieve the aim of building the 100 galleys but also save the Republic money.

The periodic reports, 1581–1713

The periodic reports in the following period provide a clear example of the diffusion of knowledge about management and the invention of a new narrative on the organizational dimension of the Arsenal. First, the galleys are presented in terms of unit costs. These costing practices were already in use in the textile industry (e.g., the *Compagnia Datini* from Prato); however, shipbuilding is a much more complex endeavor, with thousands of components, operations and workers, and with production and assembly occurring inside a vertically integrated organization.

Then, calculative practices about workforce are developed. These start "counting" humans as a production factor. On this basis, Molin in 1633 shows that the goal of 100 galleys will never be achieved, because there are not enough man months available. He observes that while it is important to have a goal, this must be realistic: therefore, he suggests setting it at 50 rather than 100 galleys.

Calculative practices are extensively presented in all periodic reports up to 1713 (Zan, 2021), in terms of measuring the reserve – and discussing the lack of progress over time – to the procurement of various kinds of materials, to organizational aspects concerning the workforce, including hiring and career procedures. Comparative cost issues are also discussed, which today we would call "costs for decision making". The organization of space is discussed in parallel to the conversation about the organization of labor.

Indeed, it is this conversation on organizing that constitutes the order of the Arsenal in the painting by Maffioletti, and to which it is indebted.

Implications for business and management historiography

In short, what emerges from the archives is a consistent set of innovations at the Arsenal toward the end of the 16th century, in terms of logistics, layout, standardization, organization of labor and accounting: in a word, "management" (or in any case, *maneggio*). Those writing about the 100-galley reserve started a new conversation about managing, at the same time inventing new concepts and metrics (in particular, the notion of work in progress and costs that started to appear in the reports around the 1630s) while applying calculative practices to all factors including workforce and developing new tools (budgets).

The innovations taking place at the Venice Arsenal in that period can be defined in a variety of ways, depending on the research perspective: it is an early case of the introduction of a factory system with thousands of workers; it demonstrates a consistent approach to administrative coordination; it is an early example of managerialism that uses the logic of efficiency, although the term itself is not yet used; it provides a logic that anticipates by some centuries the process of the economization of the world, in Polanyi's (1977) terms. The logic of efficiency, according to which, for instance, old masters should be replaced to allow control of the work team, may appear obvious today, but for Drachio, proposing it came at the cost of exile, as this questioned many of the socially accepted norms about lack of discipline in the workplace, and in rules at various levels (including the need to discipline the relationships between individual members of the aristocracy and the Arsenal as an entity).

An interesting question at this point is how all of this is reflected in the international literature on business, management and accounting history. Not that much, as it turns out, and less than in the past for what concerns management literature:

> While the lack of attention to what came "before" the industrial revolution is the subject of recent debate in business history, the management field is even more blind to this issue (indeed, the lack of any sense of historicity in this field deserves investigation on its own).
>
> (Ari & Zan, 2021, p. 2)

It is very rare, in fact, to see questions about the development of management practices over time. When the question is posed explicitly, the answers are disarming: for Ansoff (1984) everything began with the 20th century; Pfeffer (2009) refers to 60 years of management history! The rare cases that refer to the history of management usually discuss the history of management thought (with the inevitable reconstruction from Taylor, Ford, Fayol, etc.) rather than the history of practices. Anglo-American accounting historians have until recently had a colonial attitude and refer only to studies in English. Apart from a genuflection to the founding father of double-entry bookkeeping, references, periodizations, concepts and relevant authors considered inside the English-driven historiography have nothing in common with other continental traditions of accounting, nor even accounting history – including the Italian one, with a curious position that with some irony has been characterized as "After Paciolo, nothing" (Zan, 2004b, p. 257).

> There is no mention of the Venice Arsenal at the turn of the 16th century as a crucial example of the development of an early form of modern management, with two exceptions. George (1972) focuses on the history of management thought and provides a chapter on "Management during the mediaeval period", with a long summary of Lane's research about the Venice Arsenal (pp. 35–40). Peleton et al. (2018) explicitly refer to this contribution but mention neither the Arsenal nor Lane, despite the wonderful title of their book.
>
> (Ari and Zan, 2021, p. 2).

A new attention to state manufacturing can be identified in recent discussions in business history (Gelderblom and Trivellato, 2019), wherein shipbuilding could be an interesting setting for comparative analysis across space and time (Ari and Zan, 2021).

More generally, the findings at the Venice Arsenal raise serious questions about mainstream historiographies at three levels:

1   A critique of firm-centrism that characterizes economic thought and economic history – management, budgets, organization of labor and so forth are crucial elements in the history of economic development, but they are not necessarily linked (or were not borne with reference) to the firm. In our context, it is state manufacturing, and an entity that is not even conceptually "a firm", but rather a simple preindustrial production entity belonging to the state bureaucracy. The notion of profit is totally absent (no income, simply costs, and no profit as a concept).

2   A critique of Anglo-centrism – if business and economic historians, starting from Chandler himself, had read Italian documents in the archives or studied European (preindustrial) bureaucracies, perhaps they would have found a different history of the economic development of modern capitalism. And indeed, when they do or did, a completely different narrative emerges (see Lane himself, and the suspicious silence of Chandler on his contributions, although they were working in the same institution).

3   A critique of the economization of explanations "that ascribed organizational innovations and, crucially, the very creation of administrative coordination to the need to economize on costs and achieve economies of scale in new production plants" (Zan, 2005, p. 479). This, once again, ignores the role played by state bureaucracies all over the world

(including in China) and the invention of factory systems involving thousands of workers well before the US railroad construction.[3] Or even worse, it provides false historization, such as when Johnson and Kaplan (1987, p. 6) affirm: "Before the early nineteenth century, virtually all exchange transactions occurred between an owner-entrepreneur and individuals who were not a part of the organisation", a sort of John Wayne or cowboy epic economics.

From a management history point of view, the archival documents regarding management of the Arsenal at the turn of the 16th century are extraordinarily valuable. Examining them allows a re-appropriation of early episodes in the history of management, overcoming Anglo-centric (or indeed, Western-centric) views.

Moreover, from an organizational point of view, the Arsenal represents a unique opportunity for investigating patterns of organizing over a long period of time: potentially, a 900-year longitudinal study. While in the following centuries the Arsenal (and Venice in general) loses its innovative value and "competitive advantage" compared with the rest of the world, at least Europe (Rapp, 1976), it provides a context for reconstructing a "stratigraphy of organizing" (Zan, 2019b), discovering different layers of management conversations across time, in one of the most important economic sectors affecting and profoundly reflecting the whole history of Europe (see the contributions of Lane or Braudel, at this general level, though missing the lens of organizing).

Implications for heritage studies

More broadly, this also opens up a new perspective in looking at the Arsenal as a case of industrial heritage: behind "the walls" (their preservation, research, protection), the site has important intangible meanings in terms of skills, tasks, jobs, professions (the wonderful word *mestiere/mètiers* missing from English) that took place in those walls, built precisely in order to make this (now) "intangible" thing happen.

Within the intangible, the attitude relating to organizing is often forgotten, because of the lack of interest of historians in management and organizational studies, and vice versa by management scholars, toward history. On the contrary, for the Arsenal in the 16th–17th centuries, the development of management knowledge is a source of "outstanding" value with regard to the intangible meanings of the site.

Finally, the management perspective peeps out from under an additional sense in heritage studies, as a way of looking at the process of recovery, preservation or re-uses of industrial heritage. How can we revisit the whole debate of valorization of the Arsenal from a management point of view: how feasibility, project management, resources and institutional issues were involved in the discussion? This is what we will discuss in part III of the book. And, to what extent does the whole picture give an account of – and adequate representation to – the intangible value of the Arsenal in its history as a production entity? Part IV of the book will try to address these issues.

Notes

1 Actually, an old rope machine was donated a few decades ago to the Municipality by one of the last rope factories to cease operation, but this still lies in pieces – dismantled, not studied, not presented – somewhere in the Arsenal. See Chapter 7 and Figure 7.1.
2 The reader can find details and an extensive bibliography in Zan, 2004a, 2021; Zan et al., 2006; Zambon and Zan, 2007.
3 "Given the small size of firms prior to mid-nineteenth century, specialization would remain confined within the company circle. The business would be run by the proprietors, while the need for the thorough and meticulous internal organization, detailed statistics and cost-calculation methods, which were to become such a marked feature of the modern firm, was not yet felt" (Chandler, 1986, p. 20).

2   From manufacturing to industry

The turn in the late 19th century

Pasquale Ventrice

The Venice Arsenal: between artisanal and industrial

This investigation references past trends in the types of production carried out within the Venice Arsenal and reflects on more recent history as the Arsenal was transformed from an artisanal facility into an industrial factory.

The complex of buildings added to the Arsenal on the former *Zimole* islands between the 16th and 18th centuries are exceptional for their location, size and specificity of use. Reflecting on the specifics of the artisanal facility helps to define its character at the point when it was transformed into an industrial factory and offers useful clues about what to do with the Arsenal – without ignoring some of the debatable and controversial choices of recent years.

First, we must consider whether enough of the character of this vast complex survives to distinguish the area, covering about one-tenth of the whole of the historic center of Venice. This will influence decision making and ensure that any changes perpetuate that character.

It is useful to place the building typology of the Arsenal within that of the historical city and its territory, in which the constant co-mingling of terrestrial and aquatic elements gives rise to a distinct environment. This highly unusual geomorphological situation dictates the allocation of residential or industrial space within any settlement, but particularly so in the Arsenal where allocating natural and artificial areas either to industrial plant or artisanal manufacture had to be weighed.

Examining the history of work and production as it developed over the centuries around the lagoon perimeter and in the estuary is essential to identifying the limits imposed by nature and environment, and to understanding what Norberg-Schulz et al. (1992) calls the *genius loci*, referring to the unusual tensions within the spirit of the place: physical and material, notional and immaterial.

DOI: 10.4324/9781003200055-4

One of the most significant transformative events, impacting both the Arsenal and the city, was the Unification of Italy, which triggered the transformation of the industrial mechanical-metallurgical factory. The shift to metal required a fundamental transformation of the factories and buildings, whether re-purposed or built "from scratch", in order to bring ship building up to the standard of other European nations. The use of metal allowed for considerably larger hulls which in turn demanded larger spaces for their construction.

With the new production methods and the gradual increase in scale it became apparent that the settlement needed to expand beyond the city perimeter. The limits imposed by the lagoon geomorphology cannot be overcome either by art or by nature, as Galileo puts it, and precluded expansion on the scale seen in larger and less environmentally sensitive arenas.

The military arsenals of the 19th century: bridging artisanal and industrial manufacture

Italian industrial-military arsenals of the second half of the 19th century and the first decades of the 20th century, like the Venice Arsenal of the old regime, still needed a centralized workforce. Industrial manufacturing has the advantage over artisanal production in that the plant used throughout the production process, from smelting through to machining components, used the same energy sources.

Large military arsenals constituted a complex construct of tools and machines dedicated solely to the production and assembly of warships. The military-industrial shipyards specializing in the construction of warships aimed to implement up-to-date industrial labor processes, despite pressure to save costs.

The punishing economic recovery that followed Unification was hampered by the fact that, due to pronounced territorial fragmentation resulting in a failure to integrate the various regional economies, the Italian Industrial Revolution lagged quite far behind that of other European countries.

In the decade following Unification there were few patents relating to mechanical inventions.[1] Industrialization did not get underway or produce results until the last two decades of the 19th century and was confined initially to the shipbuilding industry.

As modern military ships required more specific and sophisticated technologies than the shipyards of the wooden ships era, the military yards adopted new centralized productions systems to produce steamers of small tonnage and hybrid merchant ships composed of wood and metal.

Initially, as part of the centralization process, large military-industrial shipyards aimed to concentrate production of all the necessary components of shipbuilding on site and to deliver them directly to the point of assembly, thereby reproducing the historic model – albeit with significant variations. However, the officers of the military factories were soon forced to acknowledge the advantage of outsourcing construction of components: external factories guaranteed superior quality products at lower prices, as compared with in-house production.

Outsourced production came to be a distinguishing aspect of these uniquely complex, geographically widespread organizations and became intrinsic to the shipbuilding industry.

The new industrial manufacturing plant could function even with its physical installations geographically dispersed. Privately run, dispersed specialized companies and plants proliferated, enhanced by new discoveries and innovations during the Second Industrial Revolution.

Over a couple of decades, private enterprise stimulated innovative processes widely enough to form a network of companies capable of autonomous production, illustrating how centralized, strictly vertically integrated management, as used at the Venice Arsenal, had become onerous and unworkable as a means of co-ordinating numerous steam-powered industrial plants which, freed from traditional sources of energy (wind, water) could manufacture any product.

Broadly speaking, this was the model adopted by the Admiralty for all three Italian arsenals (Venice, Taranto, La Spezia), which were restructured as industrial factories in the second half of the 19th century.[2] Essentially, they behaved as self-sufficient entities, occupying a category of production management which provided within a single, centrally coordinated production entity the various specialist departments required to produce the components of a single product: in this case, the hull with its rigging.

The concentration of the workforce in a single workplace perpetuated a characteristic of the Venetian artisanal facility, as revived in the 18th and 19th centuries, when the manpower employed in the Venice Arsenal exceeded even that of its 16th-century heyday.

At the end of the 19th century, an abundance of human resources coincided with the intensive use of machinery driven by motors in series (both hydraulic and electric), producing an environment favorable to mass production. From this point of view, the Venice Arsenal, judged in terms of the human and mechanical resources of the organization, is undoubtedly a model that influenced subsequent industrial arrangements, at least in Italy.

In general terms, heavy manufacturing industry initiated an industrialization process that could no longer be confined to the military arsenal model and which split into two types of manufacturing enterprises: one adhering to Fordist principles and the other transforming ideas into products.

Comparing artisanal and industrial manufacture: general aspects

The management of a modern industrial factory, understood as a single, coordinated system, is the result of a prolonged process of maturation. Conversely, in both the Republican period and in the post-Unification period, the processes for organizing labor and management within the Venice Arsenal displayed a markedly artisanal, that is, piecemeal, character.

Obviously, this model is only partly comparable with that of modern industry, which evolved in an era of more sophisticated mechanization and of assembly lines. The industrial enterprise we know today displays a more articulated and complex character due to significant achievements in both scientific and technological fields.

At a general level, we need to pinpoint some particular characteristics that the industry began to assume during the 19th and 20th centuries:

* Separation of producers from owners of the means of production.
* Centralization of the workforce in a single workplace; in this respect the Venice Arsenal is truly a precursor of the modern manufacturing sector.
* Intensive use of machines powered by engines in series – hydraulic, steam-powered or electric.

To illuminate the distance that separates the modern industrial factory from artisanal facilities, we should examine the phase in which it was transformed, through various incremental, intermediate steps, into a real and distinct industry. In this very interesting period tradition and innovation intertwined inextricably and took form in consumer technology products. Some of the earliest articles of this period, such as automobiles, were still the product of artisanal facilities, and though characterized by the use of serial component products, still tended to retain the distinct and exclusive characteristics of manual production.

Industrial manufacture at this time still looked to artisanal manufacture, with some products retaining characteristics of one-off craftsmanship and with the customer able to choose the producer on the basis of the quality and characteristics of the product, and on the reputation of the brand. Due to their individuality and functional specificity, such products were not widely available and were intended for a restricted elite; as a result, designers and producers could indulge in developing original design solutions which greatly increased value. In contrast, the production line allowed for quantity and lower prices, albeit at the cost of quality. In this transitional phase there was closer contact between producer and user who, exercising his rights as a customer, could request aesthetic or functional customizations.

In conclusion, at the end of the 19th century products on offer for private use (whether durable or disposable) afforded a reasonably wide and often very sophisticated range of choice without losing the individuality associated with artisanal products. It was only around the turn of the 20th century that a widespread manufacturing industry was established in Europe and the success of this massive industrialization offered perceptible economic advantages and gave rise to an economy characterized by a wide variety of supply.

The contiguity between manufacturing and industry in the naval domain

Ships built in the Venice Arsenal, both under the old system and during the decades either side of the turn of the 20th century, can be regarded as special products and, to a degree, have a unique character analogous to that of the proto-industrial manufacture, of which the most distinct example is the automotive industry of the time.

The proto-industrial factory was still considerably constrained by artisanal methods of production and standardization applied only to individual products. Modern ships were made of metallic components and mechanical devices cast in bespoke thick, wooden molds, hand-forged and machine-forged iron components, and wooden components used for interior divisions and furniture. In addition, machines were introduced and used extensively in the Arsenal. Numerous machine tools, power generators and overhead traveling cranes, or "bridge cars", were installed, disrupting the internal structure and even the layout of the *tese*.

This was followed by the disruption of all the old departments, now assigned to new functions, and by the installation of a Decauville track with railcars built to carry bulky products and heavy artillery. At the close of the 19th century blast furnaces, steam-powered rolling mills and numerous machine tools (hammers, lathes, etc.) were introduced. These were increasingly employed in both the intermediate stages of assembly and construction (specifically in making and working single pieces), and in the completion stage. Nevertheless, this method of production resembled the previous "punctiform" method, in which the physical locations allocated to construction were the covered slipways.

Concerning the working of metal as a new shipbuilding material, once cast in sheets, in the *galleazze* they underwent intermediate forging and machining consisting of the following:

1   Shaping iron sheets and iron bars in specialized machines (hammers, rolling mills).
2   Pre-forming hull parts.

3   Manually forging small components and assembling engine parts.
4   Installing armaments in warships.
5   Processing timber components.
6   Plant and instrumentation.
7   Rigging to fit out and complete the ship in all its parts.

The planning of post-Unification arsenals adhered to the idea of concentrating equipment and plant according to organizational criteria. This was partly influenced by the large arsenals of the north Atlantic and partly by the need for economy. In making the appropriate changes, managers accommodated the conditions and influences of the old artisanal shipbuilding facilities, and especially in the case of the Venice Arsenal, tried to exploit the historic functional configuration.

Notes

1   It is interesting to mention that the first legislative grant of a patent in Europe is found in an instrument of the Venetian Senate dated March 19, 1474 (A.S.Ve, Senate terra, register 7, sheet 32) which allowed for the registration and protection of "any new and ingenious device . . . refined to perfection" within the city for a period of nine years.
2   In 1865 the Naval Command of La Spezia was established, the Livorno Naval Command was dissolved, and the shipyard was ceded to private industry. Following the Battle of Lissa in July 1866, the Venetian provinces were ceded to the Kingdom of Italy. In October 1866 the Adriatic Naval Command was transferred from Ancona to Venice.

Part II
The state of the art
Protection, preservation and re-use

3   The recovery of the Arsenal

The process from 1980 until today

Claudio Menichelli

Premise

At the end of World War I, the Venice Arsenal was a fully operational ship-yard complex, providing full-time employment for about 4,000 people. Soon after, however, a first phase of decline began, culminating in the private ship-building company, CNOMV, being granted a concession over the basin area in 1932. After a hint of recovery during World War II, when the workforce swelled temporarily to 5,000, from 1955 on a second more relentless phase of decline resulted in the CNOMV concession being extended to cover the entire *Novissima* area in 1957. This act was a milestone in the fortunes of the complex. Previously always a singular, unified area, the Arsenal now became divided into two parts: military in the south, private in the north.

From the 1960s, considerations about the maintenance and regeneration of the Arsenal have made it essential to find a new use for the complex. The re-purposing of structures, and the broader issue of new uses for the entire set-tlement, were not fresh topics; from the early 14th century to the 1920s incre-mental adjustments in response to evolutions in shipbuilding had been the usual practice. By the middle of the 20th century, however, this approach was no longer viable. The realization dawned that the topic of re-using had been explored and indeed had already begun. Between the 1960s and the early 1980s, fact-finding and feasibility studies abounded, as the administration and heritage professionals attempted to define feasible routes to renovation.

Little happened on an operational level, particularly in terms of restora-tion. In the South Arsenal, the Navy struggled to meet its commitment to maintain an area now too large for its own resources. In the North Arsenal, private concerns were really only using the basins and entirely disregarding the *Novissima* buildings. As a result, a substantial part of the complex was in effect abandoned: essentially all of the *Novissima*, part of the *Galeazze*, and almost all of the southeast sector, from the *Corderie* to the *Isola delle Vergini*. The sole instance of relief during this period was the restoration of the

DOI: 10.4324/9781003200055-6

monumental Land Gate (1973–1976). Conceived and directed by the Superintendency and funded by the Dante Alighieri Society,[1] under the UNESCO Private Committees Programme for the Safeguarding of Venice, this was the first concrete sign of a renewed interest in the fate of the complex.

Steps to recovery in the South Arsenal

The 1980s marked a turning point. A series of concrete initiatives and interventions laid the ground for a consistent recovery process. The Superintendency and the Venice *Biennale*, each with its distinctive roles, worked in parallel to recover of a large part of the South Arsenal. In 1980, during the 1st Venice Architecture *Biennale*, The *Strada Novissima* opened the interior space of the *Corderie* to the public: 20 architects (Venturi, Gehry, Koolhaas, Hollein, Isozaki, Graves, Stern, Krier, Bofill, Ungers, Dardi, Purini, Anselmi, Gordon-Smith, Moore, Tygerman, Greenbereg, Scolari, Kleihues, Portzamparc), occupying ten bays on two sides of the building, created a varied and fantastical architectural stage set, interpreting the *Corderie* space. The installation occupied only 70 of the 318 meters of space, accentuating the very elongated nature of the building, but the exhibition's reception proved the Arsenal's potential as a venue for cultural activities. It also fired the starting pistol on re-purposing the building and provided the starting point from which to expand the exhibition function of the South Arsenal.

At the same time, the Superintendency launched a vast program of investigative studies exploring the recovery of the complex. A topographical survey of the entire Arsenal and a comprehensive historical study were made. In 1983, a substantial round of special funding, known as FIO (Fund for Investment and Employment), allowed the Superintendency to begin the restoration of the *Corderie*. Nearly 8,000 square meters of the roof was renovated within four years (Figure 3.1). Also, in 1983, the Superintendency oversaw the restoration of the *Porta da Terra* and the gates of the *Artiglierie* and of the *Sale d'Armi*. In the following years work went ahead inside the *Corderie*. The re-installation of wood and trachyte floors and new windows and the restoration of doors and all internal structures in metal and wood added up to a comprehensive restoration of the building.

Work carried out by the Superintendency afforded the *Biennale* a permanent home, starting in 1990 with the explosive and controversial *Aperto* section of the Venice Art *Biennale*. Since that date, the Superintendency has essentially allocated a share of the Ministry's regular spending to the Arsenal, focusing on the roofs, always the most critical element in any building restoration. Between 1990 and 2010, the interventions progressively covered the entire southeast Arsenal, involving approximately 25,000 square meters of roofing and encompassing, in addition to the *Corderie*,

Figure 3.1 Restoring roofs at the *Corderie*, 1983 (concession by Ministero della Cultura – Soprintendenza Archeologia Belle Arti e Paesaggio per il Comune di Venezia e Laguna)

the *Artiglierie*, the *Gaggiandre*, the north and south *tese* on the *Isolotto* (Figure 3.2) and parts of the coal *Tese delle Vergine* and of the *Sale d'Armi*. Funding from the British committee, the Venice in Peril Fund, ensured the preservation of the Armstrong and Mitchell hydraulic crane (Figure 3.3).

In 1999, the *Biennale* secured a concession from the State Property Agency over 50,000 square meters of the southeast area and has been gradually expanding into the restored spaces (Figure 3.4). The Superintendency's restorations contributed to supporting the *Biennale*, but since 1999 the *Biennale* has also taken an active role in preserving and re-purposing the architecture. Some of its works have incorporated the Superintendency's restorations: the plant and equipment used for the initial fitting out and a significant number of building restorations, for example, the Telemetry tower, the *di Ferro* garden, the *Tese delle Vergine*, the *Teatro piccolo* and buildings 49 and 50. In recent years, the *Biennale*'s restoration activities have picked up pace, encompassing the entire North Armory complex (Figure 3.5) and the South Armory (not yet completed). Among the works included in the *Biennale*'s re-purposing program, the *Ponte dei Pensieri* and the Italian Pavilion were particularly significant. The *Ponte* affording direct access to the *Giardino delle Vergini* from outside the Arsenal, is of great significance in terms of accessibility. The Pavilion, located in one of the two *Tese delle Vergini*, is proof of the synergy that has arisen between the

Figure 3.2 The *tese dell'Isolotto* after restoration, 2008 (photo Menichelli)

Figure 3.3 Armstrong and Mitchell crane (photo Menichelli)

Superintendency and the *Biennale*; achieved through collaborative funding, design and management of works, it was completed in 2007.

Other substantial initiatives and interventions, all begun in the 1980s, helped turn around the fortunes of the South Arsenal, in particular those by the Navy and MAV *(Magistato alle Acque di Venezia)*. In 1986, after suppression of the Navy garrison hospital, located in the *Convento di Sant'Anna di Castello*, the health services were transferred to one of the factories of the *Celestia* area. The renovation was the first of a series of numerous interventions by MAV aimed at saving and re-purposing the architecture within the military zone. The works concerned *tese* 12, 13 and 14 of the Old Arsenal, structures given over to military uses, the *caserma Marceglia*, the canteen, transferred from the *Fonderie* to the *officina dei Fabbri* on the *Stradal campagna*, and the *Squadratori* building. In 2004, the crenellations on the towers of the *Porta d'Acqua* and the wall flanking the *Rio Dell'Arsenale* were restored. Alongside these, we should note important initiatives undertaken by the Navy and restoration interventions conducted by the Genio (Ministry of Defense Engineers). The first of these was the transfer of the Institute of Naval Warfare from the Livorno command to the Venice Arsenal in 1999, at which time it was renamed the Institute for Maritime and Military Studies.

Figure 3.4 Corderie during *Biennale Architettura*, 2004 (photo Menichelli)

Figure 3.5 Sale d'Armi south, interior (photo Menichelli)

The relocation, as the change of name implies, also established an additional role for the Arsenal as the cultural hub of the Navy. The transfer was followed by a challenging plan to reorder the *Squadratori* as the headquarters of the school, but this was not funded. As part of this markedly cultural new departure for the Navy, an institutional roundtable was established (The Arsenal Project Steering Committee) to develop a project of reorganization and expansion of the National Museum of Naval History (on this see more extensively Chapter 9 and Chapter 10 of this book).

Among the interventions carried out by the Ministry of Defense Engineers, we should remember the 2008 restoration of the *Officine dei Remeri*, the Library, the Officers' Mess, the Headquarters buildings, as well as numerous maintenance works. Between 2012 and 2017, the Navy was also behind a series of significant recent interventions – proof of renewed commitment to the Arsenal's recovery on the part of the Ministry of Defense – encompassing the 15th-century wet docks of the *darsena nuova*, the 14th-century remains of the defensive walls, the *tese dell'Isolotto* in the *Piazzale Della Campanella*, one of the *Novissimetta tese* and the *Squadratori* building – allowing the Garbi Hall, an exceptional 80 × 25 meters and the largest hall in Venice, to be brought back into use.

Steps to recovery in the North Arsenal

Because private shipbuilding activity was concentrated in the dry docks and left the *tese* substantially unused, the extension of the concession over the *Novissima* area in 1957 did nothing to benefit the conservation of the buildings. Not only that, but the recovery of the South Arsenal begun in 1980 was not mirrored in the North Arsenal.

At the beginning of the 1990s, when there was as yet no hint of any recovery initiatives and the physical deterioration of the buildings was accelerating unchecked (Figure 3.6), a clear and concrete sign of possible recovery arrived with the formation of Thetis, an engineering and marine technology company, in 1993. Having been granted a concession on the buildings of the *Lamierini*, the *Modelli* and *tesa* 106, Thetis entrusted the recovery of the buildings to the architects Igino Cappai and Pietro Mainardis, who completed the works in 1996 (Figure 3.7). This intervention, of considerable quality, was an extraordinary innovation in the Arsenal landscape and illustrated a viable route to recovery within the complex. The approach chosen was to conserve the existing building shells and insert new architectural elements within them, making them suitable for new functions.

However, there was no immediate reaction to this sign from Thetis. At the end of the 1990s, the state of degradation in the North Arsenal appeared to be very serious: almost all the buildings needed extensive and demanding restoration work and some showed substantial instability and were

Figure 3.6 Tese delle Nappe, 1989 and 1999 (a: concession by Ministero della Cultura – Soprintendenza Archeologia Belle Arti e Paesaggio per il Comune di Venezia e Laguna; b: photo Menichelli)

Figure 3.7 Lamierini building, Thetis offices (photo Menichelli)

uninhabitable (Figure 3.8). In short, the North Arsenal, in entirely private hands except for the "Thetis case", was lagging seriously behind the publicly held South Arsenal – somewhat contradicting the widely held conception that private initiative is more dynamic than public.

The year 2000 marked a turning point in the recovery of the North Arsenal. Once again, it was a public initiative that unblocked the dire situation that had prevailed for 50 years. Following a series of surveys and investigations, MAV launched an impressive program of interventions. Beginning at the end of the 1990s and taking ten years to complete, it touched every building in the North Arsenal and effected a complete rehabilitation of the built environment (Figure 3.9).

Another decisive factor in the recovery of the North Arsenal, and the complex generally, was the role played by planning efforts. The overall operation, which gave rise to a truly organic project, started with the Municipality and a 1998 Ministry of Public Works Notice – PRUSST (Program of

Figure 3.8 Tesa 110 in 2000 (photo Menichelli)

Figure 3.9 Tesa 104 after restoration by MAV in 2006 (photo Menichelli)

Urban Regeneration and Sustainable Development in the Territory). From that initial operation, and successive elaborations, a sort of Master Plan was drafted, eventually approved in 2001, providing a framework for defining detailed planning choices. The next phase defined two detailed plans: one for the North Arsenal (approved in 2003) and one for the South Arsenal (never approved).

Planning initiatives defined the direction for the recovery of the Arsenal, indicating the end goal for each area, the degree to which each building was transformable, and guidelines for redevelopment processes. Four broad-brush priority goals were identified: Research Hub and Development Hub in the North Arsenal and Navy Hub and Cultural Hub in the South Arsenal. For the South Arsenal these directions essentially provided a snapshot of what had happened and what was happening in the Defense-owned area (partly Navy, partly *Biennale*); for the North Arsenal they described a developing process, based partly on the Thetis model and partly on other shipbuilding activity around the basins.

In 2003, following approval of the detailed plan for the North Arsenal, conditions necessary for regenerating the area were falling into place.

To facilitate the recovery process, also in 2003, SAV was founded – a limited liability joint-stock company owned by both the State Property Agency (51%) and the Municipality (49%). It was tasked with planning, activating and managing projects aimed at enhancing the Arsenal (see also Chapter 4 and Chapter 5 in this volume). In the same year, a programming agreement was signed between the Municipality, MAV and CNR to plan

Figure 3.10 Tesa 105 (photo Menichelli)

and effect the transfer of CNR ISMAR headquarters from *Palazzo Papa-dopoli* to the Arsenal. In 2006, SAV held four design competitions, and the Agency gave a concession over the medium dry dock, the large dry dock and six *tese della Novissima* to the *Consorzio Venezia Nuova* (CVN herein after) to house a control center and maintenance unit for MOSE.

These two operations began a definite turnaround in the fate of the North Arsenal. The CVN projects addressed four areas: (1) the tower of *Porta Nuova* – Study Center and a multipurpose structure; (2) *tesa* 105 (Figure 3.10) – North Arsenal Reception, headquarters of CVN, and a business incubator; (3) *tesa* 113 – canteen and restaurant; and (4) a new raisable bridge connecting the North and South Arsenal. The CVN concession opened up a phase of planning and interventions in the area of the basins and in *tese* 107–112.

In 2009, the Superintendency launched a new protection measure for the Arsenal, replacing the previous one of 1986. The new measure, which provided specific directions for each building and for the uncovered areas, was designed to protect and promote the process of development of the complex. It used the same designation of areas and buildings as the Master Plan and integrated perfectly with it.

At the end of the first decade of this century, after an initial period of inertia, while MAV completed the restoration of the buildings, the majority of the program's projects took off in quick succession, either through direct

Figure 3.11 Porta nuova tower interior (photo Menichelli)

Figure 3.12 Tesa 101, CNR Auditorium (photo Menichelli)

action or through CVN. In 2010, the new headquarters of CNR ISMAR (in *tese* 103 and 104) and HBB were established; 2011 saw the completion of work on the *Porta Nuova* tower (Figure 3.11) and the offices of COMAR in a former bunker; work on *tesa 105* was completed in 2012; and in 2013, the Generator Building and the new offices of CVN in *tese* 108 and 109 were finished, together with the expansion into *tese* 101 and 102 of the headquarters of CNR ISMAR (Figure 3.12).

Transfer to the Municipality of Venice

In 2012, ownership of the Arsenal, excluding those areas still in use by the Ministry of Defense, passed to the Municipality of Venice. The transfer took place on February 6, 2013, under an agreement between the Municipality and the Ministry. The transfer defined the extent of the areas to be reserved to the Navy and retained all of the existing concessions.

Since then, numerous initiatives have been undertaken to revive the Arsenal, including opening certain areas and, on several occasions, the entire complex, to the Venetian citizenry. Among all the initiatives, the most important was the drafting, in November 2014, of a new Master Plan, updating the 2001 Plan. This document was defined by the Office of the Arsenal of the

Municipality as "strategic non-normative, for the purposes of establishing a shared basis of general goals for the enhancement of the complex of the Arsenal as the new engine of sustainable development for the entire metropolitan area of Venice".

In conclusion, the steps briefly summarized here prove the great consistency of successive interventions made between the 1980s and today, interventions that have in most cases physically revitalized the entire complex, and in certain cases also changed its image.

However, there remains a clear separation between the north and south areas. Above all what stands out is the lack of an overall strategy or a general direction to systematize achievements and organize future plans. It is evident, after the intense activity that changed the face of the North Arsenal between 1994 and 2013, that intervention has slowed down. It might be that this is a period of natural adjustment, or that both public and private interest in the area is waning.

Note

1   The Dante Alighieri Society is made up of numerous committees around the world: funding for the Arsenal came from the Aarau (Switzerland) and the Quebec (Canada) committees.

4 The role of the Agenzia del Demanio (State Property Agency) in enhancing the Arsenal

Ettore Cammarata

"Nessun dorma!" ("None shall sleep!") This was the categorical imperative objective handed down to the Regional Directors of the newly created State Property Agency in 2001, when I was called to Venice as Regional Director with jurisdiction over the Veneto territory.

No real estate owned by the state shall sleep! Carefully thought-out strategies were requested for the state's properties. Starting with a census, a shrewd audit of state assets and disciplined actions designed to add value, combined with a technical and administrative audit for identifying the potential of each asset, followed by direct intervention, even re-allocation of properties to more appropriate and more marketable uses, in close collaboration with the Municipal Authorities. What better opportunity than the derelict state of the vast Arsenal complex?

At this point it was necessary to identify a single entity capable of formulating overall design hypotheses. That was a good opportunity to establish, for the first time in Italy, a joint-stock company owned by both the State and the Municipality: SAV. Shares were allocated 51% to the State Property Agency and 49% to the Municipality or, to put it another way, the Agency operated as the owner and the Municipality exercised jurisdiction on territorial development.

The opportunity was unique. Operating separately, as had been done up until that time, did not work. Had the company not been created, the State could have gone on issuing piecemeal concessions simply to produce income, without any vision of appropriate management for the real estate complex, while the Municipality would not have had any direct dialogue with the property.

There was an urgent need for the State Property Agency, working together with the Municipality, to define a shared urban planning approach: a novel approach, the first such attempt in Italy. I was immediately in favor, so I got involved.

Accordingly, SAV was incorporated to provide services in the parts of the North Arsenal already vacated by the Navy: running feasibility studies,

DOI: 10.4324/9781003200055-7

making renovations, assembling public/private finance initiatives to fund enhancements, managing property assets such as buildings and infrastructure, stimulating financial resources, coordinating management activity across the property and getting involved in the real estate market.

Who owned what in the Arsenal before 2013?

The complex was entirely state-owned or, more accurately, registered to the Agency in the guise of various departments and at different levels of ownership, within three branches (the situation would change with the 2013 new law devolving a large part of the northern area of the complex to the Municipality):

* State Property Agency, Art-historical Branch, North Arsenal (35%).
* State Property Agency, Ministry of Infrastructure and Transport Branch, barracks area, North Arsenal (5%).
* State Property Agency, Ministry of Defense Navy Branch, west and south area (60%).

The North Arsenal was not finally surrendered by the Navy until 2001, when it passed into the jurisdiction of the Agency.

Thetis, a marine technology company, occupied portions of the North Arsenal. Around the Basins there were other organizations: Palomar, Cav, an artisanal consortium "Arzanà 2000", ACTV, CVN, a rowing club and private shipbuilding companies.

The 16 *tese* (shipbuilding sheds) surrounding the *Novissima* basin were all in a derelict state: the entirely collapsed roofs of the *Galeazzo*; the clearly deteriorating walls of the *Nappe* (see Figure 3.6); invasive vegetation in the *tesa San Cristoforo*, and *tese* 100 and 102; substantial collapse of the enormous fume hood serving the two sheet metal forges in *tesa* 103, (now part of CVN), roof trusses and collapsed roofs of the *Novissima*.

It was necessary to survey the buildings and review the existing concessions in order to confirm whether they were compatible with the proposed new functions.

Enhancement by the State Property Agency, 2001–2005

The Agency assumed that an agreement between all the institutions involved in the Arsenal was needed from the outset and has operated on three fronts. It has signed an agreement with MAV and with the Municipality to begin a program of work to safeguard and restore the *Novissima* (consolidating the quays, stabilizing the roofs and exterior walls). It has participated in the Steering Committee convened by the Navy for the National Museum of Naval History project, slated for the South Arsenal, also in the preliminaries for drafting the Detailed Plan South Arsenal, following the principle of unifying the entire complex. And it has encouraged the elaboration and approval of the North Arsenal Detailed Plan in agreement with the Municipality.

Why drill down on the North Arsenal? What need for the North Arsenal Plan?

There was a pressing need to clarify allocations to new productive activities, the placement of research institutes and restoration needs of the sheds. Following from the urban planning instruments of PRUSST and the Master Plan of 2001, the Municipality needed PPAN in order to define shared guidelines.

The Municipality had adopted PRUSST as early as 1998. This was the first proposal for urban redevelopment of the complex, in which four functional macro poles were identified for several areas:

1   Research pole (north).
2   Handicraft-production pole (northeast).
3   Exhibition-cultural pole (south and east).
4   Military pole (west and south).

The Master Plan 2001 foresaw six areas of development, with two novelties: the *delle Virgini* area as a city park and the barracks area providing a public green and a new access route. The Plan, approved by the Agency in April 2003, and signed jointly with the Municipality, Veneto Region, MAV, Navy and Port Authority, has become an important tool for the subsequent implementation, providing a new blueprint of functional organization.

Particular attention was paid to the issue of external accessibility and routes within the complex, prioritizing open access of the Arsenal to the city. Until then, access to the South Arsenal was via the *Corderie*, and to the North Arsenal, either by crossing the Thetis concession or via the old walkway suspended over the water near the boundary wall. Provision was made to provide access to the *tese della Novissima* across the open area, gained by demolishing a part of the barracks. To aid internal circulation, provision was made for a new swing bridge next to the *Gaggiandre tese*, and another between *Isola delle Vergini* and the *Castello* area.[1]

Plans were made to re-organize communal infrastructure services, running on old and fragmented networks in the individual areas, including new fire protection, electricity, sewerage and computer networks.

The development of the complex, seen essentially as a place of production activities, now conformed to the guidelines on six newly defined functions specifying "intangible" and cultural activities, in each area (see fig. 6.3 on page 62):

- *Stecca Novissima* and *Galeazze* – new access, research and exhibition activities.
- *Nappe* and the *Tese di San Cristoforo* – cultural and recreational activities.
- *Tese della Novissima* – cultural activities, exhibitions and research laboratories; and *Tesa* 105, a new entrance to improve internal circulation.
- *Torre Alberaria* – reception and information point.
- Loading Bays – outdoor exhibition activities.
- Barracks – residential areas, and creation of a covered area to aid circulation and to allow access to the *tese della Novissima*.
- Dry Docks – productive activities and new artisanal activities, to include an educational element.
- *Lamierini* buildings – offices and research laboratories, capitalizing on the positive experience with Thetis.
- Creation of open spaces around the docks and quays.
- Submariners' Building – in an interesting departure, accommodation for future Arsenal employees.

In April 2002, the cost of restoration was estimated at around €400 million and the cost for urbanization and infrastructure adaptations at around €150 million.

In order to implement the project, state funding would definitely be needed; later on, the private sector could get involved. Recent legislative provision no. 410/2001 means that private sector investors could be granted long-term concessions.

I would like to highlight the state's close involvement. It contributed by committing considerable financial resources in those years. Future completion costs could be borne by individual concession holders (private or public).

The Agency's objectives in the wake of the North Arsenal Plan

These are the innovative procedures which, in my opinion, could have been used instead. Once the best urban destinations had been identified according to the North Arsenal Plan, the Agency could have put new forms of Management in place later:

- For example, by starting the tender process to select an Asset Management Company, provided for by the D.L.n. 98/2011, which is configured as a financial instrument promoting local territorial development. It enables the creation and management of a real estate investment fund, which may use public assets in the hands of the State, to redevelop them in specific ways. By offering publicly owned real estate assets as collateral against the issue of public funds, a form of public-private partnership could have been configured. This would have been an innovative solution for the State's property dealings.

* Alternatively, it might have been possible to make use of the provisions of recent law no. 410/2001 – the so-called Enhancement Fee, for the maximum duration of 50 years, favoring future concession holders directly and the Municipality indirectly. Concession holders with 50-year contracts would benefit from fees reduced by the amount of depreciation and costs incurred; the Municipality, as a partner in the redevelopment process, would be receive a payment from the State, at a rate equal to 10% of the annual rent.

There have been some positive outcomes:

* The derelict buildings were physically restored and given new purpose, though these occurred as part of a general neighborhood rehabilitation project.
* A forecast of infrastructure and urbanization was made.
* At the same time, the Agency scored highly in achieving the further objective of rationalizing government's use of real estate and reducing rental costs, a significant example of which was the allocation of some *tese* to CVN.
* The building interventions managed by SAV and financed by the Agency, and the definitively assigned concessions, were achieved in adherence to the programmatic guidelines adopted with the North Arsenal Plan, which ensued that programming and operations continued along the same lines.
* In this initial period, the State invested considerable economic resources for a recovery that was long overdue.

I can say categorically that in the Arsenal, all state-owned properties have been roused from their lethargy and drowsiness. We can say that the goal of "*Nessun Dorma*" has been won!

Note

1   At this point it was decided to postpone the construction of the first bridge until the renovation of the *Torre Alberaria* (Mast Tower), located at its head, had been completed; the second bridge was built in 2005.

5 The Società Arsenale di Venezia (SAV)

From state control to municipal control

Roberto D'Agostino

In this chapter I will try to set out why the hypothesis of returning the Arsenal to the citizens of Venice has little chance of being realized – unless we can break the deadlock we're in.

A little history

After the closure of the military yards in the mid-1960s and the transfer of the Naval Command to Ancona, the Arsenal began to suffer a very marked physical degradation which the residual shipbuilding activity, surviving into the second half of the 1990s, was not enough to mitigate.

However, during those years, two initiatives showed the way to a possible recovery of the entire complex. In 1980, the Venice Architecture *Biennale* staged an exhibition inside the *Corderie*. Twenty internationally renowned architects submitted façades for an installation which became the manifesto of Postmodern architecture. Today the Venice *Biennale* continues to enjoy an annual concession, a situation which has long impeded the formulation of long-term recovery programs. In the first half of the 1990s, the company Thetis was established: a courageous attempt to establish a center for research of maritime technology in a remote, poorly accessible location, surrounded by buildings on the point of collapse.

Such initiatives, presented as experimental and temporary, while important, do not make a scalable plan of recovery for a heritage complex still substantially in the hands of the Navy and the State Property Agency. The Arsenal remains remote and cut off from the city.

In the second half of the 1990s, the Municipality gained the capacity to implement programming and project management. Armed with experience won in other "problem areas" of the city, for the first time it began to take a hand in the life of the Arsenal.

The opportunity arose specifically from the need to respond to an employment emergency resulting from a crisis in the shipbuilding activities within

DOI: 10.4324/9781003200055-8

the yard and the failure of the company that managed them. The company that had inherited Fincantieri's shipbuilding concession, covering the whole of the demilitarized North Arsenal, had initiated bankruptcy proceedings. Productive activities were increasingly rare in Venice, which risked becoming a tourism monoculture, so the Municipality, in particular Mayor Cacciari, assumed an active role in trying to keep such activities alive.

At that time, the Municipality was active on many fronts: it supported shipbuilding activities in the Arsenal by installing the maintenance company ACTV, it put moral pressure on CVN by getting involved with some of its activities within the Arsenal, it contracted with MAV on some of that authority's annual program of works around the complex and it pushed for settling CVN within the Arsenal.

Finally, in 2001, it developed a real Master Plan, approved almost unanimously by the Municipality's Council, for the organic re-use of the entire Arsenal, followed by operational urban plans that provided the transformative rules for interventions, and which became the founding document on which future development programs will be based.

The Master Plan is a programming tool with a strong political content. It is the first indication of general objectives, the operational actors, the main uses to which different parts of the complex will be put and the specific actions to be implemented in order to achieve these objectives. It is a tool that has no time limits but that can be periodically adapted to the conditions that arise throughout the recovery process.

The final act in the city's process of adopting the cause of the Arsenal occurred in 2003, when the Municipality and the State Property Agency formed a joint-stock company, SAV, owned 51% by the State Property Agency and 49% by the Municipality.

Since then – apart from rhetorical declarations on the importance of the Arsenal, its potential to function as an engine of the city, its indivisible unity, the need to return its use to the citizenry – there has been a stand-off between the (very thin) forces of those who wanted to follow up on these statements (chiefly SAV and a slew of citizen associations), and those who wanted to continue to use the Arsenal for their own particular purposes, always presented as being of a public and institutional nature.

The actors in play

The Arsenal has long been divided between three powers that have agreed between themselves how to apportion their individual parts: the Navy, the *Biennale*, the CVN. The Municipality, representing the citizenry to whom the area should have been returned ought to have favored, promoted and regulated the recovery, acting through SAV, without needing to take a firm political hand to keep the ambitions of those three powers within the framework of the declared shared objectives.

The Navy's presence in the Arsenal is not up for discussion. It currently allocates space to important military studies activity; however, it excludes the city from huge spaces which it does not use. The Navy reduced its operational presence by relocating shipbuilding activities to La Spezia and the Naval Command to Ancona and subsequently conceded numerous buildings to the *Biennale*. Despite this gradual contraction, the Navy retains vast areas of open water and buildings of which it uses only a small part.

The *Biennale* has privatized all the accessible parts of the south and east, and controls huge spaces, for profit, without effective oversight by the Municipality. Those portions of the *Arsenale* assigned to the *Biennale* represent a considerable tract of urban space – buildings, streets, squares, quays –visitable only for a fee at certain times and closed off entirely for half the year. The perception that the *Biennale* is in some way "high culture" prevents people from fully grasping the affront to the citizenry that this situation represents.

CVN and its companies have tried to seize the entire northern part, exerting its weight on all public and private entities, even putting the area to illegitimate uses at odds with those purposes agreed with the state – activities halted only by the legal proceedings linked to the entire MOSE operation. CVN made continuous requests for far more than was functionally necessary to protect the lagoon. These requests were supported by MAV, which should theoretically have been exercising public oversight. The requests had two objectives: to make spaces available for non-institutional activities, as for example occurred when one of the CVN companies built, inside the Arsenal, components of a degassing platform to be used in the Upper Adriatic, that is, nothing to do with safeguarding Venice; and, second, to guarantee an ongoing concession over spaces and buildings in which to manage MOSE, thereby gaining an advantage at the time when the management contract should have been put out to international tender.

The only body that acted in the interests of the citizens by sticking closely to public programs approved and defined by the Master Plan was SAV. The municipal administration initially gave arms-length support, but then, often "distracted" with respect to the aims of CVN, turned openly hostile.

Since MAV did not exercise its oversight functions but in fact colluded with the illegal activities of CVN (as found by judicial review), SAV was the only public body located physically within the Arsenal to defend public interests. In reaction to this, attempts were made to delegitimize and to suppress SAV and Mayor Orsoni eventually made the decision to dissolve it.

SAV, though operating under the conditions described, has spent over €10 million on restoration interventions, paid for out of non-ordinary funds, without impacting municipal budgets. It has opened up the only accessible spaces to date: the *Porta Nuova* tower, and the *tese di San Cristoforo, delle Nappe* and others.

At the same time, it has drawn up a vast project for the recovery of the entire North Arsenal (now languishing in some drawer of the Municipality), has made the Arsenal part of important European projects and, above all, has tried to prevent its privatization, thus engaging, without real support from the public, in unequal conflict with powerful opponents.

The statute creating SAV anticipated outcomes beyond the mere enhancement of the Arsenal and included the recovery of other public assets conferred by the shareholders.

Once SAV was legally incorporated (about which more later), studies and proposals put together in previous years, as well as projects approved meanwhile by the State Property Agency, were collated and organized within a single multi-year Business Plan. Included within these were projects to recover and re-purpose historic buildings, adding up to a cost of approximately €80 million, such as the construction of 70 social housing units in the former *Celestia* convent; fitting out large event and concert spaces in *Galeazze*; the creation of accommodation (residential, students, and others, provided for in the Master Plan) within the former Submariners' Building; a center for traditional shipbuilding (comprising a yard, a school and a museum showcasing a reconstructed *Bucintoro* and the recovery of the traditional watercraft); and many more besides.

The Plan apportioned space for each of the projects. More significantly, it set out the financial plan, indicating sources of funding – project finance, European funds, private funds and self-financing with Arsenal-generated income – as well as a timescale for implementation and associated cash flows.

None of this would have been of significance in the history of the Arsenal but for lucky coincidence and interactions of three actors: a director of the State Property Agency, Stefano Scalera, a true civil servant, who realized the absurdity of keeping Arsenal real estate within state hands and considered Venice to be its rightful owner (see Chapter 4 in this volume); a minister who proved to be in favor, Minister Grilli; and a determined Board of Directors serving SAV.

These factors favored the inclusion, in the Public Spending Review Act of July 6, 2012, article 95, for transferring ownership of the entire Arsenal to the Municipality, with conditions particularly beneficial to the city: entrusting preservation, restoration and development to CVN; the constraint that all resources produced within the Arsenal area must be re-invested in the Arsenal; the inalienability of all the parts of the complex; and retention by the Navy of only those parts that are functional to its institutional activities.[1]

As was expected, all those parties that had thought they could treat the Arsenal as they pleased reacted violently and managed to insert, within a successive Bill, clauses recovering many of their privileges and extending to CVN rights previously reserved to the Navy. (Growth Act 2.0, introduced by Corrado Passera, Minister of Infrastructure and Transport 2011–2013)

The Municipality, rather than hindering this reaction, actively encouraged it, playing an active role in driving those claims. Despite forceful protest by citizens and associations who felt the city had been assaulted by these changes, the Municipality intervened at the final reading to have the Passera Bill modified in ways that were even more detrimental to recovery and entirely in line with the interests of private entities.

The amendment by CVN to the Passera Bill, achieved with the help of their ministerial contacts, was limited to extending to CVN the same privileges that the Act of Transfer had extended to the Navy. As the Bill was passing into law, two new amendments were submitted. The first, signed by Senator Giaretta, simply deleted the article in favor of CVN. The second, drafted with the cooperation of Mayor Orsoni and signed into law by Senator Casson, radically modified the text of the Transfer Act, upholding the right of CVN parties to make extensive private use of the complex. Moreover, SAV was suppressed.

The Growth Act 2.0 of December 17, 2012, article 221, reads: "Paragraph 19-a is replaced by the following: ownership of the compendium constituting the Venice Arsenal, excluding the portions used by the Ministry of Defense for its specific institutional tasks . . . is transferred without fee, in its current physical and legal state, to the Municipality, which ensures its inalienability, enhancement, recovery and redevelopment. To this end, the Municipality guarantees:

1   free use, for the portions of the Arsenal used by MOSE for its operational centre and ancillary services, . . . for the operational life of the MOSE system;
2   free use, for the activities of the Biennale di Venezia foundation, by virtue of the nature of the functions discharged by that body, the CNR and in any case by all public entities currently allocated space to perform institutional functions."

By apparently conveying the asset to the Municipality, but simultaneously legally removing from its control areas for the Navy, CVN and the *Biennale*, and suppressing the managing company SAV, the conveyance lost most of its meaning. It prevented the Municipality from implementing the previously agreed wide-ranging policies and undermined the possibility of managing the entire complex as a single whole.

Subsequent events saw the dissolution of SAV and its replacement with an Arsenal Office within the Municipality, with no power, later dissolved.

Only the much publicized legal proceedings in which CVN became embroiled impeded a project to transfer the entire North Arsenal to CVN.

Despite the detrimental changes desired by the Municipality at that time, the legal transfer of the real estate package provided that:

- The areas ceded to CVN be limited to those strictly necessary to its maintenance activities.
- All the *tese* south of the Great Basin remain at the disposal of the Navy until the materials contained in them be transferred elsewhere.
- Income produced by the use of the Arsenal be ploughed back into the recovery and preservation of the Arsenal itself.

Regarding the commitment by the Municipality, to establish a body dedicated to the management, recovery and enhancement in place of the dissolved SAV, a small digression should be allowed. The commitment had been made both publicly before various citizens associations, and during the Municipality meeting, to dissolve SAV.

To get an idea of what we are talking about, the Arsenal covers 47 hectares, two more than Vatican City, three fewer than the Villette of Paris; it has a covered area, on solid ground, of about 200,000 square meters, which is three times the size of the palaces of Caserta or Versailles. To recover the Arsenal and open it to the city in its entirety requires making programs, projects and contracts; finding funds; and interacting with different bodies at an international level: it is clear that all of this can be achieved only by a dedicated and competent body. In devising a body that is up to the task, surely we envisage something like the entities deployed throughout the world to tackle matters of this nature and complexity and not a marginal municipal office, subordinate to dozens of other decision-making centers.

A body of that type would certainly be unacceptable to those receiving rents from the Arsenal: unacceptable to Municipality/CVN relations to cede the whole of the North Arsenal to CVN; unacceptable to the local Naval Command, which inappropriately occupies a huge amount of space, not allowing access to or movement through the whole of the Arsenal; unacceptable to the *Biennale*, which has a monopoly on the use of various areas, legitimately in some cases, not so in others, and which impedes access to Venetians to more than half of the entire complex.

Since the Arsenal was transferred from the State to the Municipality, nothing provided for by the Transfer Act or announced by the city's governors has been realized:

- Nobody has asked CVN or the Commissioners for a business plan or an assessment of how much space they justifiably need.
- No one has asked the Navy to hand over the *tese* in the South Arsenal, let alone to cede a right of way on the west side of the *Bacino Grande*.

- Not one single euro has been spent on the Municipality's initiatives for recovery of the Arsenal in years.
- Not one dedicated body has ever been constituted. The Arsenal Office too was dissolved, without the decision being in any way justified to the city.
- The city kow-tows to the *Biennale*, but no serious request has ever been made to open its assigned spaces to the population or to find uses for those spaces in the periods when the *Biennale* is not using them.

There is no program, no project, no deployment of capabilities that could lead us to hope that the Arsenal will step up to the levels that the city requires.

Yet the plans for the Arsenal are there, with necessary resources detailed in the business plan for the recovery of the entire North Arsenal, identifying procedures, subjects, funds and timescales for the recovery, as drawn up by the now-defunct SAV.

Similarly, various associations, chiefly the Association for the Return of the Arsenal to the City (ARAC), have put forward numerous interesting and viable proposals to make large parts of the Arsenal immediately usable and liveable. Evidence of this can be found in documents from the Arsenal Forum, detailing a series of proposals re-purposing areas for activities related to Venetian culture and production.

The problem is that these proposals do not get a readership because the intrigues of the powers I have mentioned are effective in preventing the Municipality from listening to voices in the city and making progress toward the recovery of the Arsenal, drawing up a strategic project and setting up an adequate implementation structure.

The previous provides some clarity on events surrounding the Arsenal – a clarity that I fear justifies my pessimism about its recovery and its restoration to use by the city as a whole.

Note

1 "Urgent provisions for a review of public spending, with invariance of services to citizens", to quote article 19-a: "on account of its historical and environmental characteristics, ownership of the compendium constituting the Venice Arsenal, with the exception of the portions used by the Ministry of Defence for its specific institutional tasks, is transferred to the Municipality, which ensures the inalienability, indivisibility, and the enhancement of the Arsenal by delegating its management and development to the *Società Arsenale di Venezia spa*. . . . The sums received for the compendium's use are to be used solely for the management and for the development of the Arsenal by the above company. . . . For the purposes of this article, the State Property Agency, in consultation with the Ministry of Defence, proceeds to the delimitation and demarcation of the compendium and the delivery of the same to the company *Società Arsenale di Venezia spa*."

6 Some comments from a management studies perspective

Luca Zan

The chapters in this second part of the book constitute an important starting point for an overall reflection on the interventions in the recent history of the Arsenal. Some basic elements are confirmed: the extension of spaces and their constitutive complexity (paradoxically: the abundance of potentially usable historical spaces, of a huge size) and the considerable financial resources that this will require, all taking place within a context of institutional fragmentation. These aspects are no surprise for those with even a passing knowledge of the change processes at the Arsenal in recent decades. However, some less intuitive aspects emerge from these chapters.

(a) Much has been done in terms of recovery, yet its scale remains unknown. The three chapters qualitatively describe how in the last decades, starting from 1980 as a symbolic date, the processes of recovery at the Arsenal have been intense. Over the decades, the Superintendency has played an almost heroic role; other institutional actors have done a lot. It is important to understand the effort made, and the underlying logic, in order to give an idea of the overall size of the intervention.

Strangely enough, however, it is difficult to reconstruct the amounts of financial resources and effort that various institutions have put in place. Paradoxically, while usually sponsors, funders and donors try to leverage image and legitimacy from their investments in the recovery of cultural heritage, in this case, aggregate data are lacking and their reconstruction is extremely difficult. A few important studies are now available, in particular the contribution by Gambelli (2017), which with great difficulty provides some data. The lack of systematic data is itself the result of the overall state of affairs: linked on the one hand to the institutional fragmentation affecting the Arsenal in recent decades (which persists and is perhaps getting worse) and on the other to the reporting methods used by individual institutions in a situation characterized by an overall lack of accountability.

DOI: 10.4324/9781003200055-9

(b) A question of governance, with two related aspects. There is the issue of a deficit of transparency, democracy and leadership (alas, these phenomena are not new to Venice: see Samiolo, 2012, on the MOSE event). The needs of creativity (and shared views) that the design of re-uses of the Arsenal would require, given the constitutive complexities discussed earlier, are not adequately taken into account.[1]

In this regard, legal reforms present a situation of persistent fragmentation and legislative cunning. The institutional fragmentation of the Arsenal toward the end of the 1990s remains unresolved today. On the contrary, the complex situation regarding concessions is now ratified by law, making dialogue between various stakeholders extremely complex. Law 221/2012 sanctions the transfer from the State Property Agency to the Municipality of Venice "in the state of law and fact in which it is found", undermining the process of negotiation between actors within a potentially unitary framework. The whole affair is disconcerting in its procedural aspects: the regulation of structures, uses and stakeholders is completely outside any systematic planning practice. Law 221/2012 is the final step in a series of subterfuges and expedient moves in which various actors confront each other, almost "by stealth", in conceptually improper venues: first the *Decreto Milleproroghe* (hard to translate itself: "A law of thousand deferments", putting together several issues that need urgent legal acts, without any unifying logic, and providing room for last-minute secret lobbying) and then the Passera Decree on digitalization (*sic!*) and Law 221 itself, created by who knows whom and how. A few words in this or that document upset the institutional conditions and strategic feasibility of the whole project of re-using the Arsenal (as D'Agostino reminds us in Chapter 5). Figure 6.1 and Figure 6.2 describe the conflict between ownership and concessions, and thus greatly limiting alternative re-uses.

The whole event deserves "genealogical" reconstruction in a report to the international community. Overall design, feasibility, responsibility and accountability remain foreign concepts in this procedure. It is difficult to explain to an outsider the existence of the so-called area *sine die* (*sine die* means "without any precise deadline"), which the Navy has to vacate because it is no longer using it for institutional purposes but whose times and modalities are formally *unregulated* by law (rather than simply being unregulated).[2] This makes any planning process even more difficult: a *sine die* plan is a contradiction in terms, absolving all possible actors in a situation of complete non-responsibility.

(c) The incompleteness of the Master Plan, which has now vanished. It should be stressed that the 2001 version of the Master Plan – unanimously approved by all parties in the town council – clearly defined a crucial aspect of the Arsenal's re-use strategy, namely the allocation of areas for naval

Figure 6.1 Ownership representation in the Master Plan (Città di Venezia, 2014, Table 7, p. 19)

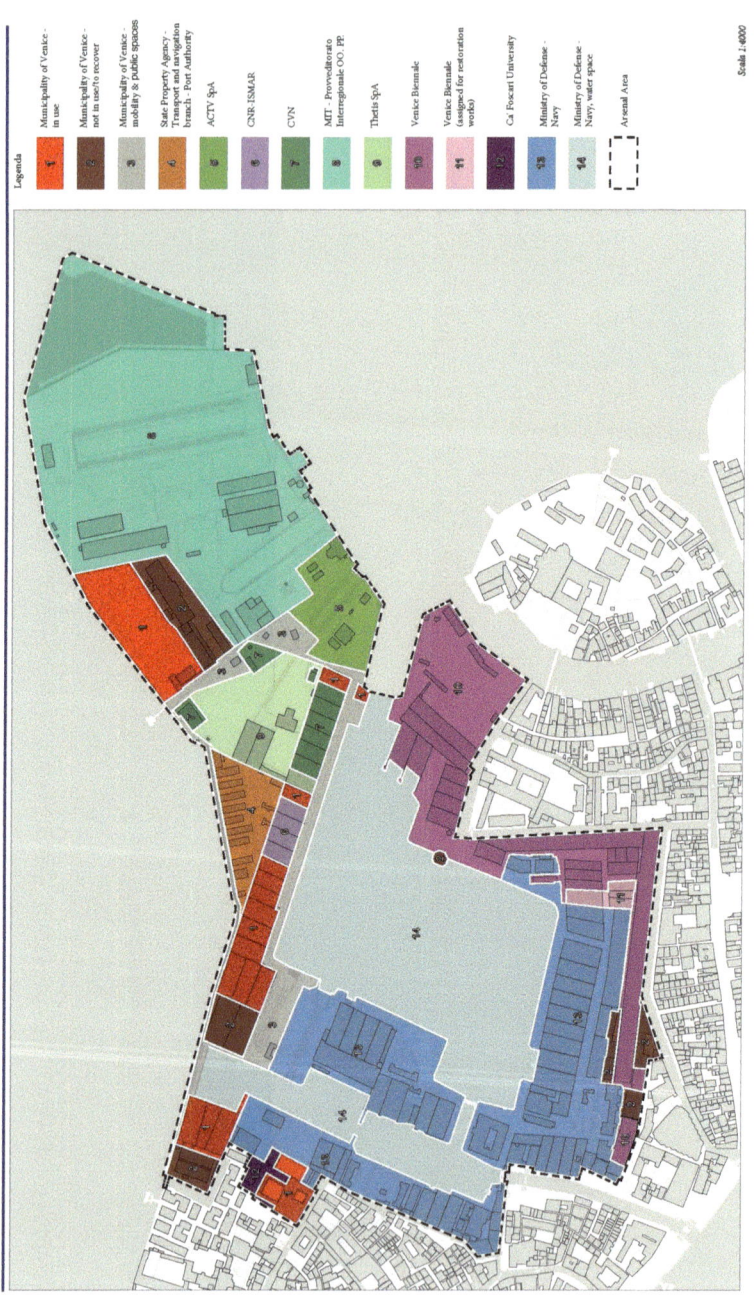

Figure 6.2 Current concessions and uses in the Master Plan (Città di Venezia, 2014, Table 9, p. 21)

production and research (the northern part of the Arsenal), culture and the Navy (the southern part). The next version in 2014 further defined six main areas within the Arsenal, specialized for different activities (Figure 6.3):

1   Innovation (green) – main functions: research laboratories; offices; educational, cultural and recreational equipment; craft workshops. Additional functions: shops; bar and restaurants.
2   Collective services (orange) – main functions: collective housing; hospitality; cultural, educational and recreational equipment; shops; craft and research laboratories. Additional functions: bar and restaurants; offices.
3   Manufacturing and boating (yellow) – main functions: manufacturing; services for nautical activities. Additional functions: cultural and recreational equipment.
4   Culture (purple) – main functions: cultural and recreational equipment. Additional functions: shops; bar and restaurants.
5   Housing (red) – main functions: housing, cultural and educational equipment. Additional functions: shops; bar and restaurants.
6   The Navy (blue).

The overall *concept* is certainly important, and ultimately convincing, particularly in terms of a plurality of re-uses, made possible (or indeed, made necessary) by the huge spaces involved. As mentioned in previous chapters, the Innovation area was already implemented with the establishment of Thetis and CNR, plus additional projects in this direction.

A few critical aspects, however, can be addressed, such as the following:

•   The lack of delineation of other actors – in addition to the *Biennale* – who could participate in the "cultural pole" (indeed, the *Biennale*'s very aggressive policy tends to treat the whole cultural pole as its own space).
•   The mysterious disappearance of any idea of the Arsenal Museum developed between the first and second version of the Master Plan (something we will return to extensively in part III of the book).
•   The overlooking of a central issue for any heritage site or indeed any public space, that is to say the issue of access by citizens (also an aspect we will see in part III of this book), which was never detailed, except some interesting experiments in April 2014 of "Open Arsenal" (Marsala, 2014).

Despite these (serious) limitations, this could have been an important step for the recovery of the Arsenal had there been interventions to articulate this

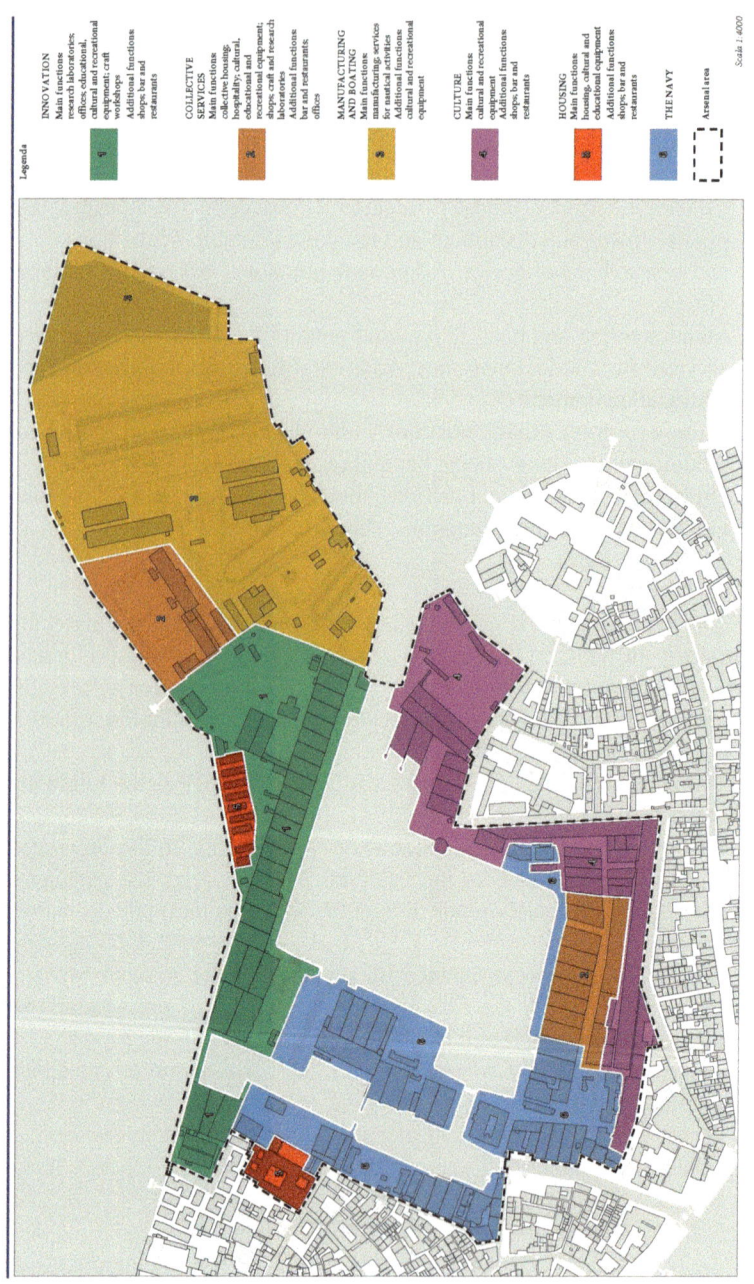

Figure 6.3 Project's contents: re-uses per area in the Master Plan (Città di Venezia, 2014, Table 15, p. 65)

macro decision into guidelines and rules, and detail individual projects and acts – which did not happen.

(d) The total absence of management and sustainability cultures. On the one hand, there is a lack of any accountability in relation to the Arsenal. Thus, it is not surprising that there is no formal, explicit and uniform logic of planning and reporting. There may be partial plans, budgets and reports within the various institutions, but they are simply not public and accessible. This in itself explains the difficulties in "giving an account" of what has been done, even when much has been done, see point (a).

On the other hand, one could argue the logic of management is marginal to this picture. Beyond the lack of access to internal (almost secret) documents belonging to different institutions, it seems there was a limited (if any) role for management sustainability within them, in terms of both investments and future running operations. As far as investments and extraordinary measures are concerned, there is no trace of sustainability analysis or examination of the financial viability of alternative solutions.

How could things have been done differently? The basic impression is that many of the interventions suffer from a strong temporal lag. They were born and conceived before the Global Financial Crisis and did not have to reckon with the context of reduced public resources today. There was no process to put forward cheaper alternatives for future investments that could have been selected. As for future management, the ambiguous boundaries of an abstract notion concerning the "Arsenal" complex, and the lack of an effective governance structure, confine the question of overall management to a very dangerous institutional limbo.

Indeed, the lack of attention to management issues is one of the most worrying shortcomings in the whole debate about the Arsenal. This topic is unavoidable in the future. There needs to be an adequate definition of budgets, plans, time frames (without *sine die*, please), priorities and responsibilities, taking into account the magnitude of projects and resources involved, with strong leadership accompanied by a redefinition of institutional arrangements. Once these limitations are removed, an immense effort of creativity and imagination remains.

Notes

1 In the survey on the difficult reconversion of ten western cities in the face of deindustrialization processes conducted by Tarr (2016), a common fact that stands out is the need for strong leadership: an aspect denied "by law" in this context.
2 "In setting out the operational arrangements for the implementation of the transfer of ownership, the Law finally ties the actual transfer of ownership to the handover of various spaces, subject to an agreement of the perimeters of the areas between the parties involved. In fact, the delivery of the areas took place on 6 February 2013,

since then identified as the date of transfer of ownership. On that occasion, the Municipality of Venice, the State Property Agency and the Italian Navy agreed on the identification of the area and the signing of the "Perimeter and Delimitation Record", which contains the planimetric representation of the new ownership of the Arsenal, with the definition of the areas transferred to the Municipality of Venice, the areas to be transferred to the Municipality but maintained in free use '*sine die*' to the Italian Navy, and finally the areas that remain in State ownership, in use by the Ministry of Defense-Marine". Master Plan, 2014, p. 16.

Part III

A controversial issue

Museum and open access

7   Toward a Museum of the Arsenal

Home of Venetian civilization, history and shipbuilding

Camillo Tonini

The ongoing debate over what form the Venice Arsenal Museum might take, its purpose, its content and its relevance requires us to think about what the museum's objectives might be and what themes would it address and explore? At the same time, it would be appropriate to understand the artifacts and thematic materials available in the stores of various municipal Institutions, which could enhance the result of this project.

Assuming that a large part of the display will be delivered via media and recognizing that for a project like this we do not need to settle for what we have had to date, it is necessary to compile a catalogue of objects that can evoke historical themes while examining the consistency of existing public and private collections. This will allow identification of core materials for exhibitions, using narratives to illustrate the political, maritime, architectural and urban evolution of the City of Venice.

In addition to the debate about the financial sustainability of the museum, there is value too in taking the opposite tack, allowing the strong thematic collections already available to suggest ways to pinpoint the varied segments of a complex journey.

Two striking examples: first, a museum within the Arsenal was initially conceived in the second half of the 18th century by Domenico Gasperoni, who collected and arranged suits of armor, martial trophies, small arms and heavy artillery pieces of various ages, which were displayed according to the taste of the time, in the doorways and entrance halls of palaces of the nobility. The exhibition spaces of this proto-museum were housed in the *Sale d'Armi* and the adjacent uncovered spaces, as described in Giovanni Casoni's *Guide to the Arsenal of Venice* (1829) and as they appear in several prints of the time collected and described by Gasperoni himself in his manuscript *Artiglieria Veneta*, of which several examples are known. On the basis of these contemporaneous testimonies, given that much of the material included in that historical exhibition remains on site, and that the ancient setting survives largely intact, it should not be difficult to revive the setting

DOI: 10.4324/9781003200055-11

to serve as exhibition space and the hub of the new museum. Those pieces originally present in the museum but now lost, plundered at the fall of the Republic, provide a further subject for study because they directly illustrate the fate of the Arsenal during the foreign occupations of the 19th century.

Second example: a few years ago, the Directorate of the Venetian Civic Museums acquired the last preindustrial rope-making machine in operation within the *Giudecca* – the property of the family firm Inio Ropes – now stored in complete disorder within the military zone of the Arsenal

Figure 7.1 Rope-spinning machine, donated by the Inio company, 19th century (photo Guastadisegni)

Figure 7.1 Continued

(Figure 7.1). In this case too, displaying this remarkable machinery, not unlike that used at the time of the *Serenissima*, could contextualize an important step in the complex process of fitting out a ship, combining it with period images and virtual reconstructions – as was in fact done in the *Corderie* for an exhibition a few years ago.]

These two examples, to which others can be added, are sufficient to illustrate the type of work proposed here. In this new Arsenal Museum, it would be possible to include collections exploring historic themes that have been limited, if not in fact omitted, from the civic museum's programs; this is partly due to limited exhibition space but also because historical displays are often displaced by temporary exhibitions of modern and contemporary art of greater interest to tourists.

There are two collections forming part of the *Museo Correr*: one, the *Museo del Risorgimento*, was long ago moved to the second floor of the *Procuratie Nuove*. The other, the *Sale di Francesco Morosini*, has recently been restored, but it is not large enough to accommodate the vast quantity of historical materials from the palace of a Venetian captain in *Campo San Stefano*, which today is either in storage or dispersed on loan to public institutions.[1]

Another case is that of the Doge's Apartment in the *Palazzo Ducale*: its wall and ceiling decorations have been covered up for safety and its historical furnishings removed, all to accommodate temporary art exhibitions aspiring to have a showcase in the palace symbolizing Venetian civilization.

Here it is also worth mentioning the collections that have recently formed the heart of certain exhibitions about Venetian civilization. Those relating to the ancient shipbuilding and military complex could be revived within the Arsenal – for example, exhibitions that have already been mounted and which used the vast heritage of geographical, nautical and fortification-design knowledge preserved in various Venetian libraries and archives, such as the State Archive and the *Museo Correr* (see AA.VV., 1986; Biadene, 1990; Tonini and Lucchi, 2001).[2] To these could be added other similarly validated and soundly scientific works dealing with the publication and production of music in Venice.

The ultimate objective of this puzzle made up of established collections, analysis of different construction methods, the city's diversity both above and below water, the viability of the economy on which the power of Venice and others grew and so forth is to fuse them into a *Museo della Città* – a museum in which these various themes can be presented in captivating ways using objects, both real and skillfully reconstructed, to illuminate this challenging and fascinating capital of culture. In Venice, the most extraordinary and famous Western city of its time, such a museum is conspicuously lacking.

Notes

1   To mark 400 years since the birth of Francesco Morosini, three exhibitions have been mounted recently in Venice, at the Correr Museum; at Palazzo Mocenigo in San Polo; and at the State Archive. These are collected in a catalogue on Morosini (Buratti, 2019).

2   See also https://correr.visitmuve.it/it/mostre/archivio-mostre/navigare-e-descri Cousins/2001/12/4103/Project-11/.

8 The Arsenal Museum
Issues of spaces availability and accessibility

Franco Mancuso

The arguments for an Arsenal Museum

More books have been published on the Venice Arsenal in the last 30 years than on any other part of Venice. Many have a notably historical character or describe particular aspects, thereby providing an invaluable intellectual contribution to its complex reality. More than a few clearly have an agenda: they imagine new functions, propose organizational models, anticipate scenarios and configurations, pre-empt projects, sometimes even prefiguring the architecture, and in more than one case, address both the theme of the museum and the question of accessibility. Looking at them all together, what is immediately apparent is that the idea of the museum comes in and out of focus sporadically. In view at one moment – in the many studies by the IUAV for example – and then out of sight the next, re-emerging occasionally and then sinking away again without any apparent reason. In the Venice Arsenal Master Plan (2015 edition) for example, a key political act by the Municipality, the Museum is not even mentioned.

On one occasion, however, the debate on the future of the Arsenal seemed to be re-emerging – peremptorily, but unfortunately not very effectively. In 2002, Insula Ltd. published the deliberations of a *Study Day on the Establishment of a Museum of Archaeology, History and Naval Ethnography in the Venice Arsenal*, held in January 2002 in the Arsenal Old History Library (Caniato et al., 2002). The foreword included a document, hitherto rarely considered, entitled significantly *Arsenal Maritime Civilisation Project*. The document was heavily endorsed by the Council of District 1 of the City of Venice (comparatively quite energetic at that time) and unanimously endorsed around the city.

This document, developed by a coordinating committee,[1] was originally presented in December 2000 at the X International Symposium on Boat and Ship Archaeology (Ca' Foscari University). It was in effect a *programmatic manifesto*, very well structured, emphasizing the need and the urgency of settling a Museum of Archaeology, History and Naval Ethnography within

DOI: 10.4324/9781003200055-12

the Venice Arsenal "to be realized in agreement and in cooperation with the Naval Command and the other competent state bodies". It proposed that the museum be divided into four sections, the last of which would have been dedicated to preservation and restoration, with a dedicated school-laboratory.

At the start of the century therefore it seemed that a viable route to an Arsenal Museum was being proposed, and it was assumed that it would probably become fact, not least because the Navy seemed to be on board; at a conference in May 2002 Admiral Paolo Pagnottella, Commander of the Institute of Naval Studies, made an authoritative and much appreciated contribution when he published an article in the Insula's quarterly *Quaderni* with the unequivocal title "The Arsenal in the Future of Venice. The Navy's Project".

This intervention presaged concrete action: a few years later, the Navy promoted an agreement between the Arsenal's principal institutional stakeholders. A team of authoritative experts undertook a feasibility study of a museum, with the goal of reconceiving the Naval History Museum, housed at the time in the nearby former granary of *San Biagio*:

> A re-designed museum, open to the city, in dialogue with other examples of arsenals around the Mediterranean . . . with testimonies and documentation arranged along a visitor route exploring direct historical resonances in spaces where the functions which we study and re-create today, were once actively productive.
>
> (ISMM, 2007, p. 2)

Arguments for the museum, which would have impacted a large portion of the South Arsenal, were reinforced when the Arsenal Study Center (2006) published its own research report into the Venice Arsenal Museum of Maritime Culture and Civilization. The project was presented in February 2007 but, despite high expectations, nothing happened and all discussion on the topic ceased.

The theme of accessibility, conversely, is almost always present in the abundant flow of studies and proposals. Observations on the historical impenetrability of the encircling stone wall give rise to proposals to penetrate it with gates, passages, excavated tunnels, cut-and-cover tunnels, canals and bridges. Proposals frequently meet with opposition from those citing the need to safeguard the irreplaceable, highly fragile environmental quality which that very same wall has helped preserve – chiefly the absence of wave motion, traffic, and lights reflecting on the water. Gradually the insistence on accessibility receives less emphasis, until it all but falls away.

This dichotomy illustrates clearly that there has never been an obvious or significant link between the two themes of museum and accessibility – or at least no convincing argument to support either one as a necessary condition for the other. Therefore we must build an argument from scratch, without ignoring recent changes in both the Arsenal and in Venice as a whole.

The changing Arsenal

The pattern of ownership throughout the recent history of Arsenal has changed radically: once the Upper Adriatic Naval Command moved from the Arsenal to Ancona in 1979, the Ministry of Defense gradually reduced its military activities and its garrison in Venice. The downsizing was hidden somewhat when the Institute of Naval Warfare transferred from Livorno to Venice in 1999, becoming in effect the cultural hub of the Marina and simultaneously assuming the less aggressive title of Institute of Naval Studies.

The most significant event, however, was the acquisition by the Municipality of the greater part of the Arsenal, under an agreement with the Ministry of Defense (2012, finalized 2013) 27 hectares of the built environment were transferred to the city, leaving a mere eight hectares to the Navy. (The total area of the compound, including aquatic areas still held by the Navy, is 48 hectares.) However, we need to consider that the Municipality is not actually free to use all of its property because the agreement obliges the Municipality to uphold existing concessions granted by the State Property Agency to the *Biennale* Foundation, which has occupied a large part of the South Arsenal (for exhibitions) and areas to the east (open spaces and buildings) since 1980, both during the months of the *Biennale* proper (May to November) and also during the set-up period (all other months). In addition, under the same aforementioned agreement, CVN has the right to occupy the entire, vast northeastern sector, including the two major dry docks, for functions related to the management of MOSE.

In short, the Municipality has, in the northern sector, the *Torre di Porta Nuova* and half of the *tese della Novissima* where it hosts events of various types, mainly exhibitions and recreational events, managed since 2013 by Vela. The Navy owns and controls the wide expanse of water known as the *Darsena Grande*, part of the southern sector, and the entire western sector, which it uses chiefly for educational training and cultural purposes.

It is only in the annex to the northwest, around *Celestia*, that the Arsenal fortuitously accommodates the only occupants with a close link to the historical purpose of the Arsenal: the traditional boats of the Franciscan Rowing Club, which enjoys the privilege of overlooking the only mirror-smooth water in Venice, where the traditions of Venetian rowing can be upheld without the handicap of the choppy conditions that prevail everywhere else.

Meanwhile, accessibility within the Arsenal has changed: the centuries-old inviolability of the wall has in fact been broken on more than one occasion. A gap was first opened in 1963 along the northern stretch to allow ANCL Line 5 that circumnavigated Venice to pass through (Figure 8.1). A second opening in the wall granted general access to the *tese della Novissima*, when ACTV built a new station called *Bacini* and pushed the circumnavigation route eastward, instead of traversing the Arsenal. In 2008,

Figure 8.1 The public motorboat passage through the northern wall (photo Mancuso)

a further significant breach was opened on the eastern wall to reconstruct the bridge over the *Rio delle Vergini* connecting the *Giardino delle Virgini* in the *Castello* district – brought within the Arsenal precinct in the mid-19th century after the demolition of the synonymous church and convent (Figure 8.2). And last, the old gates in the southern wall, opening onto the *Campo della Tana*, were reopened to allow access to the area of the *Biennale* and to ancient factories on the Arsenal quay that are now part of the Naval History Museum.

Changing Venice

The Arsenal therefore has changed, but Venice too has changed over the same period, in its social and urban structure, in the activities it hosts, in its relations with the mainland and with the lagoon. Venice today is known to be experiencing negative phenomena: its residents, mostly elderly, now

Figure 8.2 The new bridge over the *Rio delle Vergini* (photo Mancuso)

number only a little over 50,000 (though reduced numbers have not pre-vented them from pushing forcefully for the Arsenal to be opened up as part of the city). Younger residents are becoming rarer, and college students almost all travel in from the mainland.

There are of course increasing numbers of tourists – 30 million per year – and their presence has a devastating effect on the social fabric of the city: houses formerly inhabited by locals have been converted for short lets, and purpose-built accommodation is now found in every part of the city; tour-ists invade the squares, the streets, the quays; they swamp the shops, and hamper good-neighborliness; they physically wear away the stones, misuse the buildings and open spaces and by traveling around in water taxis, motor boats and the armada of cruise ships, create the immense swells that erode the canals and reservoirs.

There is no point discussing this only in relation to the Arsenal: mapping the most significant tourism indicator – the lattice of tourist accommodation offered for rent through Airbnb (Figure 8.3) – clearly shows that it has now spread like a persistent mold across the entire city, breached *Castello*, and reached *Sant'Elena*. But, behold! It has not entered the Arsenal! The walls, even if no longer inviolable, are preventing an influx that can only be termed an invasion (Figure 8.4). Even so, in certain unpleasant circumstances, they

Figure 8.3 Mapping Airbnb in Venice, 2018 (photo Mancuso)

are still threatened by infiltration from unsightly and inappropriate events – carnival parties and dinners, fake weddings, costume parades, advertising hoardings and temporary exhibitions around the *tese* in the northeastern sector – events which the Municipality are usually behind.

But there are also positive developments in relation to the new museum project. The lagoon, laced with haphazardly excavated channels and reservoirs, occasionally offers surprises. The criss-crossing motorized traffic causes devastating erosion, day in day out, to the margins of the remaining salt marshes, potentially revealing, layer by layer, a wealth of only recently unimaginable archaeology. There are many examples: the sudden appearance, in the summers of 1996 and 1997, of the remains of two virtually intact vessels, a *rascona* and a *galea*, dated to between the end of the 13th century and the beginning of the 14th century and submerged for centuries in the vicinity of the vanished island of *San Marco* in *Boccalama* in the south of the lagoon; the research findings of Ernesto Canal (2013); the constant and effective efforts of archaeologists and volunteers of the *Archeoclub*, engaged in the recovery and enhancement of the two *Lazzaretti* to house finds of significance to the ancient civilizations of the lagoon; increased activity by a volunteer force that is not only citizen-based; and the tireless activities of groups and associations associated with the Forum

Figure 8.4 The eastern wall *(Rio della Tana)* (photo Mancuso)

for the Future of the Arsenal. This Forum proposes a museum on diverse themes, not solely of the sea, which is backed up by a convincing plan for accessibility and an equally convincing plan for the economic activities to be incorporated within it.

A future for the museum idea

All things considered, conditions for rekindling an interest in the idea and feasibility of a Museum of Maritime Culture seem today to be in place. So too do conditions hindering its realization, which must be overcome.

How and where can the Arsenal house it? Why now? Which museum should we be talking about? In this respect, the image projected so evocatively by Valeriano Pastor *(Arsenal and/is a museum)* at the aforementioned International Symposium in 2000 is particularly relevant. An image that the author has returned to and re-considered constantly over the course of his involvement with the Arsenal project over many years (Pastor, 2017) fits neatly inside the International Council of Museums definition of a museum and is still worth reading today:

> The museum is a permanent non-profit institution open to the public and acting in the service of and for the development of society, which conducts research on the tangible artefacts and intangible evidence of mankind and its environment, which it acquires, conserves, exhibits for purposes of study, education and enjoyment.

To which MiBACT later added "promotes by raising awareness among the public".

So let us consider the Arsenal as both a research topic and a place of study. It exhibits documents, personal objects, machines, ships, finds, technologies, within all environments (and spaces where these same elements were conceived, developed, manufactured and tested). It continues to accumulate new evidence of the rich culture of the Venetian lagoon, equipped as it is to receive the inexhaustible historic deposits from the city, the lagoon and the city's wider maritime and terrestrial domain. It records and studies them and, relying on its priceless historical and cartographic documentary archive, authenticates them. It has the technical skills, cultural knowledge and essential facilities to preserve them. It investigates and bears witness to the reach of the Arsenal in its aquatic/terrestrial duality. Its reach encompasses everywhere that supplied raw materials for the construction and maintenance of the ships: wood from the forests, in the foothills and on the mountains (and the means to get it to Venice); metals from the mines of Bergamo and of the Agordino for weapons, anchors and chains; agricultural products for sails, ropes and for provisioning the ships' crews; and the staging posts along the eastern routes, the arsenals (little more than warehouses) at Zadar and Korčula and, gradually larger the farther they are from Venice, on the island of Corfu, and in the ports of Chania and Heraklion on Crete.

It is a museum which constantly turns over the inexhaustible subject of Venice, connecting with the many museums and research organizations around the world that mull over similar issues in as many contexts. And it must be a publicly accessible institution. This is the real issue: the accessibility of the Arsenal that houses the museum. Accessibility is an indispensable condition if you wish to create practical and readable spaces and to display the eloquence of structures such as the *Gaggiandre*, the shipbuilding sheds, the slipways, the *Porta Nuova* tower, the technological equipment, the power plant, the *Nappe*, the Armstrong Crane, the historic walls, wooden roof trusses, massive columns, docksides.

It is therefore necessary to grant access from the south: for example, from the Arsenal Quay – also known as quay *della Madonna* and typically used to access the *Biennale* – through to the *Padiglione delle Navi* (Figure 8.5 and Figure 8.6), integrated with the new museum, where the first thing you would see would be a grouping of priceless exhibits, in particular the *rascona* and the galley of *Boccalama* (sizing 23 meters × 60 meters and 38 meters × 5 meters respectively, absolutely suited to the dimensions of those indoor spaces).

Figure 8.5 One of the *tese dell'Arsenale Nuovo*, inside the *sine die* area (photo Mancuso)

Figure 8.6 Ship gallery of the Naval Historical Museum (photo Mancuso)

A twofold agreement would be needed: with the *Biennale* Foundation, to permit the route just described, and with the Navy, to permit a route along the eastern front of the *Darsana Grande*. This would need to be open year-round, consistent with exhibition demands; an internal route accessible both from the south and from the *Giardino delle Vergini*, potentially linked to the existing route from the north. These routes all lie on municipal property; therefore, any such agreement would be within the spirit of the conveyance of the Arsenal to the Municipality, honoring previous arrangements, and also ensuring "the user-friendly public accessibility to the open-air areas . . . to maintain interaction and open sightlines across spaces within the Arsenal". From the museum's perspective, free circulation would be key, allowing readability of the internal exhibition spaces, essential to an understanding of the complexity of production methods once practiced in the Arsenal, not least within *Corderie*, the *Sale d'Armi*, the *Artiglierie*, the *Tese del Carbone*. Ticketing systems to suit different visitor pattens could be adapted where applicable.

This would provide enough space to accommodate our museum. Admittedly it encompasses spaces currently dedicated to art *(Biennale)* and to

Figure 8.7 Citizens and ships at the "Open Arsenal" event, spring 2015 (photo Mancuso)

higher education (in the Navy zone) but also spaces for activities linked to the presence and traditions of the sea, for the smaller shipyard (and the larger Prince of Piedmont Basin),[2] preserving them for training and research – activities that fit effectively with those of the museum.

Characterized like this, the Arsenal could provide an exceptional node linking the remarkable occurrences and myriad places of the lagoon, the city and its territories. For example, the archaeological and museological activism of the *Archaeoclub* in Venice is taking steps to form a Museum of Maritime Culture, based on the *Lazzaretti*. The city showed unanimous enthusiasm for the Open Arsenal event held a few years ago (Figure 8.7), a museum network eloquently illustrating the city's links with mainland locations connected historically to the fortune of the Arsenal, such as the mines of Agordino, the foundries of Bergamo, the woods of Montello and Cansiglio, river-borne logging and more. There is also a surrounding university network: the master's degree in naval architecture that we hope can be restarted at the IUAV, the Faculty of Naval Engineering in Trieste, the History-Humanities faculties at Padua and Ca' Foscari. And the now really impressive network of historical-naval museums all around the world.

Notes

1 The Committee to Establish a National Museum of Archaeology, History and Naval Ethnography within the Venice Arsenal.

2 Such access arrangements would now be possible. CVN recently announced a decision to move maintenance operations for MOSE flood defense barriers to Marghera. Only activities related to monitoring and managing the defenses will remain within the Arsenal.

9 The National Museum of Naval History
A reconstruction of the project

Claudio Menichelli

An important archaeological discovery in the lagoon of Venice in 1996 sparked a lot of interest, at national and international levels, in the role of Venice and its Arsenal in the history of commercial and naval shipbuilding. Two 14th-century vessels, a *rascona* (river cargo boat with sails) and a galley (a ship with sails and oars), came to light in the course of work in the lagoon commissioned by MAV and carried out by CVN under the guidance of the Superintendency (Canal, 1978; CVN, 2002). The discovery was made in the San Marco area of the lagoon island of *Boccalama* that became submerged in the 14th century. The two boats had been deliberately filled with ballast and sunk to reinforce the eroding shoreline against encroaching sea water. The unique historical event allowed us to uncover two substantially intact medieval boats, which had been lying on the seabed of the lagoon for more than six centuries and which could potentially reveal the "secrets" of shipbuilding techniques employed in the Venice Arsenal (Figure 9.1).

After the discovery there was a lengthy discussion about the possibility of restoring the two ships inside the Arsenal, creating a laboratory specializing in the conservation of waterlogged timber, which could also be a knowledge hub in the field of historical shipbuilding and maritime activities.

Not long after, in 1999, the Arsenal assumed a new and important function when the Institute of Naval Warfare transferred from Livorno to Venice and was renamed the Institute of Naval Studies (refer to Chapter 3 of this volume). In 2000, Rear Admiral Paolo Pagnottella was appointed as head of the Institute and of the Arsenal complex. He set to work immediately, seriously committed to creating a world-class Navy Museum within the Arsenal.

The Admiral's project rapidly took concrete form. In 2002, the Arsenal Project Steering Committee *(Comitato Intesa Progetto Arsenale)* was established, with a protocol signed by the Navy, the State Property Agency, the Superintendency, the Municipality, MIT and others. It began work immediately, in cooperation with MAV and the Committee for the Ancient Arsenal of Venice.

DOI: 10.4324/9781003200055-13

The starting point of the project was to reorganize the existing *Museo Storico Navale*, located in the *San Biagio* granary and in the *Officine dei Remi*, expanding it to extend into the Arsenal, within the buildings along *Stradal Campagna* (Figure 9.2): a street in the South Arsenal, lined on one side by a series of *tese*, transformed over time into warehouses and workshops and, along the full length of the other side, by the former *Officine dei Congegnatori-aggiustatori*.

In order to develop the initiative a suitably qualified group was formed, named *Progetto Arsenale*, coordinated by Captain Cristiano Patrese; within that group, a scientific Committee was established, chaired by Prof. Mario Dalla Costa and composed of the heads of component bodies: technicians,

Figure 9.1 Venice. Archaeological excavation of the wreckage of San Marco in *Boccalama*, Venice lagoon – Nausicaa Archive, dry-docking operations and survey of the wreckage of the (a) galley and the (b) rascona (concession by Ministero della Cultura – Soprintendenza Archeologia Belle Arti e Paesaggio per il Comune di Venezia e Laguna)

Figure 9.1 Continued

scholars and industry experts. The commission set up a smaller study group and an executive design group responsible for drafting the project.[1]

The objectives of the project were ambitious, but some of them were within reach and the conditions and tools required to achieve the others were in place. Briefly, they were as follows:

- To form a museum that interacts with other examples of museums of shipbuilding and navies existing elsewhere around the Mediterranean.
- To create visitor routes through the buildings, allowing materials to be viewed and testimonies examined in the spaces relevant to them.
- To create visitor routes inside and outside the Arsenal, to give a flavor of the complex in its entirety.

Figure 9.2 Stradal Campagna: *Congegnatori-aggiustatori* building (L) and two 15th-century sheds (R) (photo Menichelli)

- To incorporate the historical artifacts of the *Museo Storico Navale* in *San Biagio* with other recently retrieved archaeological artifacts representing maritime heritage.
- To create a center, linked to the Marciana National Library and the Venice State Archive, for the research and acquisition of historical, documentary material relating to maritime culture, in manuscript, in print or in archival form.
- To create a laboratory and a research center for the conservation of waterlogged timber, operationally connected to the recovery and restoration of naval archaeological finds.

Of equal importance to these objectives were those aims of the project which would have had a direct impact on the Arsenal complex and on Venice as a whole. In particular, the implementation of the museum project would have contributed to re-unifying the Arsenal, opening it up to the city and guaranteeing permeability and visibility of the settlement as a whole while promoting the preservation of the complex. With regard to this last point, establishing a museum could have guaranteed the integrity and values of the complex over the long term: the new use was compatible with the character of the spaces and the execution was strongly oriented to preservation.

The project provided for a museum floor area of approximately 19,000 square meters, in addition to the roughly 8,000 square meters of space already available in the *San Biagio* Granary and *Officine dei Remi*. As a result of the sharing arrangements introduced in 2008, these latter would have had to remain available as exhibition spaces, and this fact would have needed be accommodated within the museum's overall program (Figure 9.3 and Figure 9.4).

The entrance to the museum would have been on the *Campo della Tana*, specifically in the *Fonderie*, which would house ticketing, visitor orientation and a permanent exhibition on the history of the Arsenal. From there, visitors would pass through temporary exhibitions in the *Magazzini del ferro* and then visit the permanent exhibitions, which would have been arranged piecemeal, starting in the *Congegnatori-aggiustatori* building, and then following a continuous sinusoidal path through the 15th-century docks of the *Stradal Campagna* (Figure 9.2). The impressive 15th-century *Congegnatori-aggiustatori* building would have accommodated the "long ships" section, displaying the *Boccalama* galley. The 14th-century workshops would have hosted the "round ships" section. Further sections would have recounted one-by-one the steps of piecing together a ship: sourcing timber, the workings in the docks, metallurgical production, navigation in the modern age, naval shipbuilding of the 19th century and the advent of motor-propelled iron hulls (Figure 9.5). At the end of the exhibition route,

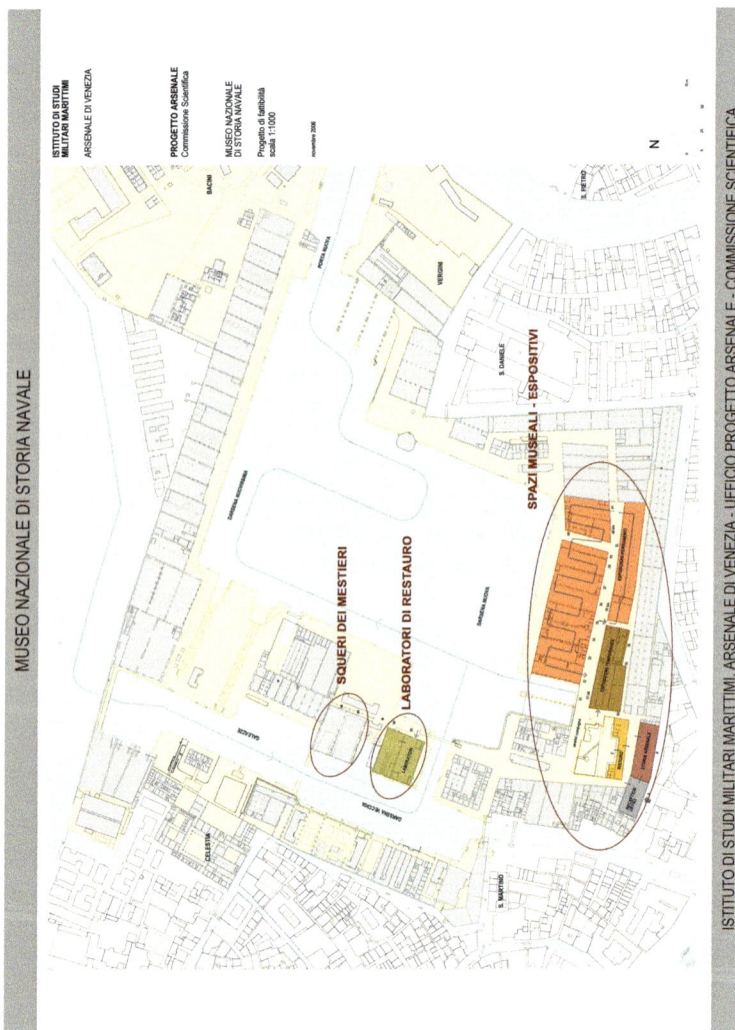

Figure 9.3 The National Museum of Naval History project: general map (photo Studio Della Costa)

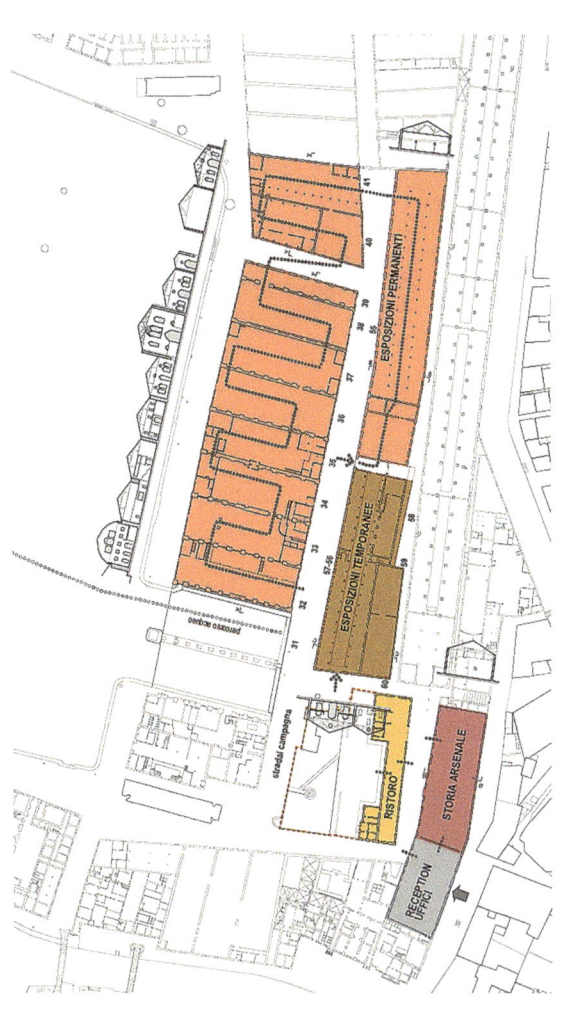

Figure 9.4 The National Museum of Naval History project: detailed map (photo Studio Della Costa)

Figure 9.5 16th-century *tese* (photo Menichelli)

or as an independent activity, one would have made a complete tour of the Arsenal, starting from the 15th-century wet docks, sailing in the mirror-like waters of the basins and visiting the 16th-century *Mestiere* Docks – spaces that have remained unchanged over time and still reflect the original dimensions of the dry docks and the wet docks (Figure 9.5).

The project was developed through the feasibility study stage, including detailed schedules for each building, a plant study, a general definition of the restoration works, overall costings and plans for managing the interventions.

The project was presented in December 2006 in the historic library of the Arsenal and steps for its implementation began immediately. In 2008, the Ministry of Defense published a European call for the construction and management of the museum. However, no tender was submitted and, unfortunately, the project was shelved and later abandoned.

The Arsenal Museum issue has never been addressed again. Since then, the institutions involved have been silent on the subject, despite prompts from academia, in the form of well-developed thesis, the first of which came from the University Institute of Architecture, Venice (IUAV) project (Perdomi, 2008; Agnoletto and Pasqualetto, 2016; Cabianca and Pellizzari, 2016).

Note

1 The Commission was composed of Mario Dalla Costa (President), Giorgio Bellavitis, Ennio Concina, Corrado Ferulli, Marco Filippi, Antonio Foscari, Raul Guastadisegni, Claudio Menichelli, Giovanna Nepi Sciré, Valeriano Pastor, Mario Piana, Maurizio Rispoli, Piercarlo Romagnoni and William Zanelli. The commission was assisted by consultants Carlo Beltrame, Luigi Fozzati and Pasquale Ventrice. The design team consisted of Mario Dalla Costa (Coordinator), Carlo Beltrame, Marco Filippi, Claudio Menichelli, Mauro Rispoli, Piercarlo Romagnoni and Guglielmo Zanelli.

10 Scenarios for the Arsenal Museum

Back to management

Luca Zan

The reconstruction of the overall debate on the Arsenal Museum allows us to identify gaps and opacities that have occurred across time and to hopefully interrupt the process of "forgetting by doing" that characterizes the discussion.

For reasons of space, only three recent projects will be examined here. All of these emerged in the early 2000s: the "Arsenale e/è Museo" project (Caniato and Fumagalli, 2002), the pre-feasibility study of CSA and CNR (CSA, 2006) and the project for the National Museum of Naval History, coordinated by the Navy under a largely participative process (ISMM, 2007). Menichelli talks about the latter in Chapter 9, with regard to the technical-professional aspects, and Mancuso discusses the former in Chapter 8.

This chapter will briefly review the three documents from the perspective of cultural heritage management, with particular attention to their museological aspects: the museum concept, the characterization of collections involved and the relationship with the pre-existing Naval Historical Museum.[1] This will lead into the analysis of more general managerial aspects relating to assumptions about visitors, overall investment and funding and current operating and governance conditions.

National Museum of Naval Archaeology, History and Ethnography ("Arsenale e/è Museum")

The monographic issue of the journal *Insula Quaderni* (Caniato et al., 2002) is an interesting example of collective reflection, almost a sort of brainstorming with several voices on a variety of aspects. It is a set of "elaborations around an idea, precise even if complex and still to be fully declined, about the use of a significant part of the Venetian Arsenal" (Zanetti, 2002, p. 5). It provides proposal from "the bottom", involving the Venice District Council 1 and a promoting Committee joined by 39 bodies: many grassroots associations but also the same Superintendency, the Universities of Venice and Padua and a couple of international bodies (listed on p. 9 of the document).

DOI: 10.4324/9781003200055-14

Three strong ideas emerge from the very first pages: the Arsenal as a museum of itself, "a new great national museum, which can refer to the history of navigation and to the Mediterranean naval culture" (Zanetti, 2002, p. 5), together with the idea of "complete accessibility and visibility of the whole area" (Castelli, 2002, p. 3); the idea of the mixture of "production" and "educational" functions (Castelli, 2002, p. 3); and the need for a common direction among actors involved. Acknowledging the original impulse of the Superintendency for the Veneto region in 1999 with the Arsenal of the Water Civilization Project (Caniato, 2002), the manifesto for the National Museum of Naval Archaeology, History and Ethnography was attached, highlighting the sections of the museum (archaeology; history of the Arsenal; ethnography, with a floating and didactic component; and preservation).

These initial aspects alone make the document valuable, and they set out the basic principles around which future discussions and controversies will revolve. Twenty-five contributions follow, on a variety of truly stimulating aspects, and which the reader can easily find almost 20 years later.

Two aspects in particular deserve attention. On the one hand, the many players share an idea of openness, with the words of Admiral Pagnottella being particularly significant:

> The new system absolutely does no longer involve architectural barriers, physical barriers between military activities and the movement of the population, because there is no reason for them. We don't hold weapons, we don't have any particular secret to keep, so the Arsenal can very well become a place of peaceful coexistence between military and civilian activities. Briefly, then, this is what we mean by a "cultural center" for the Navy.
>
> (Pagnottella, 2002, p. 33)

On the other hand, it is the relationship between the new museum being set up and the existing one that begins to be addressed, with reference to the Navy's "program . . . to dismantle the current Naval Historical Museum" (Caniato, 2002, p. 10), confirmed again by Admiral Pagnottella's words.[2]

For the management expert, there is nonetheless something ironic about this document. Guest contributor Ole Crumlin-Pedersen's account of the experience of the Roskilde Maritime Centre states clearly: "It is therefore essential to start with a master plan" (Crumlin-Pedersen, 2002, p. 17). But the monographic issue of *Insula*, although important, is not a Master Plan. Of the 25 contributions, there is not a single one that examines management problems; there is no reflection on priorities/compatibilities/compromises, and much less any consideration of resources and needs (not even in purely spatial terms). Most disturbingly, there are no numbers (square meters, let alone financial resources), nor adequate reflections on the problem of governance, beyond a climate of convergence between institutions that seem to be caught between their boundaries, which will be lost in more recent times.

Museum of the Culture and Civilization of the Sea (CSA and CNR, 2006)

Created as part of a large five-year national project on Cultural Heritage by CNR *(Progetto Finalizzato per i Beni Culturali)*, the study was elaborated by CSA and CNR and published in June 2006 (in version 3.0).

The very structure of the report – five chapters – displays the influence of an economy of culture approach:

- Chapter 1 (24 pages) starts with an analysis of the potential market for the museum, formulating assumptions about visitors per year (500,000 at the end of the first ten years). Beyond substantive data and some naïveté in the choice of comparative elements, the chapter is interesting in methodological terms. It explicitly states that the hypothesis of the number of visitors "is indispensable to be able to formulate a preliminary Business Plan on the costs and revenues of the Museum itself" (p. 27).
- Chapter 2 (21 pages) constitutes what is normally the first chapter of a museological project, with the definition of the museum concept. The basic idea is "to reconstruct Venetian history based on the development of scientific knowledge, the hydraulic tradition and the science and techniques related to the culture and civilization of water" (p. 29). Within a managerial jargon (from mission to edutainment, to product offer, etc.), the contents of the museum are then articulated into five sections: (1) science and technique in the Venetian and Veneto tradition; (2) the Arsenal and its transformations, building system and defense techniques; (3) shipbuilding techniques, archaeological history and ethnography; (4) historical heritage of technique and industry; and (5) the hydraulic tradition with annexed library. The architectural implications of the museum, mentioned in the seventh paragraph, are less clear. The definition of the architectural structure is resolved with the presentation of six low-definition images that are difficult to read, linked to the Arsenale spa website. There is a very general definition of spaces: more or less located in the area of *Stradal Campagna*. In the absence of a proper architectural plan, there is talk of an area of 8,972 square meters on two floors.[3] No further design specifications are defined for the layout, nor is there any precise definition of collections to be used, giving the impression the project is at a very preliminary stage. Explicit reference is made to a sustained use of temporary exhibitions, shows, concerts and conferences.

- Chapter 3 (36 pages), the largest part of the document, examines six major European museums and sites of the most varied nature. It is an interesting analysis, but its relevance to the Arsenal is not always clear (e.g., Tate?), and its implications are not taken up in any way. Curiosities: one of the comparative elements is the Naval Historical Museum, which in some discussions would instead be part of the project (obviously not in this one).
- Chapter 4 (eight pages) presents the business plan. Despite some oddities, it is the only project that provides costs and revenues among the various projects for the Arsenal Museum. In addition to the costs of restoration (about €10.2 million, with an unclear calculation between the costs of renovation and an unspecified "cost of equipment") and fit out (€9.4 million, including cleaning services and personnel costs not specified in more detail), the operating costs for the first ten years were determined. Rather than a hypothesis of operating at full capacity and the determination of the start-up effort of the first years, a progressive approach is used (including adjustment for inflation), according to which a break-even point would be reached from the fifth year, with approximately €5.8 million in costs and €5.6 million in revenue (€3.2 million for exhibition tickets, €1.5 million for the rent of exhibition areas, €113,000 for the auditorium, €32,000 for the rent of congress and conference rooms, €125,000 for the rent of cafeterias, canteens and bars), and then generate substantial profits in the following years (from €730 million in the sixth year to €1,660 million in the tenth year). However, this figure does not include the cost of promotion, which was "deliberately left empty because it is too variable". The lack of explanation for the hypotheses makes it impossible to evaluate the estimate of costs and therefore the reliability of the entire financial year (the exclusion of promotional expenses). In the break-even year, the hypothesis is to get 530,000 visitors, with a net ticket price of €6.6; in a questionable way, the same number of visitors is then "inflated" by 10% a year in the following period.
- Chapter 5 (eight pages) describes criticalities (viability and transport) and interventions at the Arsenal, a summary of ongoing activities by other bodies and an illustration of the finalized CNR Cultural Heritage project.
- Then follows a chapter of self-presentation by the CSA, the bibliography and a curious appendix with a short list of European amusement parks (*sic*!).

Serious perplexities emerge in reading the plan, starting from its structure (with its central emphasis on the "museum market", 60 pages compared with 21 for the museological project, and eight for the business plan).

Moreover, what is striking is the opacity of the underlying hypotheses of space uses and investments, running costs and revenues. From a conceptual point of view, two aspects characterize this project with to the others: on the one hand, the strong orientation toward "activities", with the idea of a rotation of three or four exhibitions a year. It would be interesting to check the consistency of these hypotheses with the numbers involved, in terms of costs, visitors and revenues. In any case, this emphasis on activities seems central. On the other hand, the museum focus emphasizes the evolution of the past two centuries of the Arsenal and is less attentive to its history during *La Serenissima*.

Two basic questions emerge, for which there seem to be no answer:

1   The likelihood of reaching the break-even point – even when taking for granted costs and "other revenues" (not tickets), is it plausible to expect 530,000 visitors for a museum of this size, in these (inadequately defined) spaces and with this character? And, moreover, in a situation of lack of any form of integration with the pre-existing Naval Historical Museum, so that the two museums would remain distinct and separate (with separate tickets and costs)?

2   The financial backers and the governance structure – the document is quite naïve: the idea was to rely on SAV, defined as the promoting body, but it seems that the accounts have been made "without the landlord", who in fact would never accept the idea.

National Museum of Naval History (ISMM, 2007)

The third project is extremely interesting, despite not having a happy ending. A case to be well understood, also in view of possible reopening of perspective, in its strengths but also in some intrinsic weaknesses.

The plan was drawn up by the Italian Navy's ISMM, based on years of intense interinstitutional and interdisciplinary work by an ad hoc structure, the Arsenal Project Steering Committee, that also provided "a thorough critical assessment of the historical, material, political-economic and productive knowledge of the Arsenal in its centuries-old activity" (p. 1).

It is structured in ten sections:

•   General report (four pages) – it summarizes the distinguishing features of the project: the functionalization of the southern part of the Arsenal with the objective of "respecting the unitary character of the Arsenal, open to the city" (p. 3); with a function of research and production of maritime-historical culture; in connection with neighboring areas: the

Naval Historical Museum and the planned Naval Archaeology Laboratory within the same Arsenal.

- Historical notes (three pages) – it provides a synthetic description of the salient features of the evolution of the Arsenal over its entire 900-year life span.
- Arsenal history section and methodological aspects (six pages) – this defines the main historical stages on which the narrative of the Arsenal should be structured: (a) The Medieval Arsenal, (b) The long Renaissance of the Arsenal, (c) A difficult transition: the 17th and 18th centuries, (d) Decay and reconstruction, (e) The Arsenal of the Habsburg Navy, (f) the Arsenal of the Italian Navy.
- Technical report (four pages) – it describes the restoration of the 13 buildings involved, illustrating the criteria for the interventions in terms of works (doors and windows, skylights, flooring, scaffolding, provisional works and means of work, masonry, roofing), as well as systems (electrical, lighting, smoke detection) and foundations. It also summarizes the overall cost (€57 million), with the clarification that this does not include set-up costs and services aimed at museum activities.
- Design guidelines: sections, themes and exhibition materials (seven pages) – the individual museum sections, their contents and the spaces occupied are precisely defined.
- List of exhibition materials (ten pages) – with great analytical precision, the various exhibition materials are listed in each of the sections highlighted in the previous point.
- Calculations of surfaces, heights and volumes of the buildings destined for the museum (20 pages A3 format) – the single buildings are illustrated in detail with their destination and their position in the overall map of the exhibition route according to the project.
- Guidance document on plant technologies (five pages) – the intervention logics are detailed with reference to (a) level of environmental and service quality, (b) level of centralization of plant systems and (c) integration between the building and the plant components.
- The recovery of the buildings of the *Stradal Campagna* and estimated costs (46 pages) – taking up a series of analyses developed by the Superintendency, the sheets summarizing the interventions on individual buildings are presented, with a description of the works and an estimate of costs.
- Guidance document on running costs, broad estimate (two pages) – a page of text accompanies two tables that show the expected costs that total about €2.4 million. Among these are outsourced costs for surveillance of about €1.2 million; €650,000 for electricity, air conditioning

and gas; costs for internal personnel of €162,500; maintenance costs of €75,000. No activity costs are foreseen. There is no estimated revenue, nor are there any assumptions regarding the number of visitors. The overall impression is of a radical underestimation of costs (for a museum of this size); in the case of personnel, in addition to an underestimation of the job positions (seven people, including the director), the values are equal to the total amount of *salaries* (moreover with a low estimate, for example €22,000 for the director) and not the *cost of labor* (which in Italy differs substantially, due to the so-called tax wedge).

In summary, this presents an enormous amount of work, including detailed analysis and planning of great analytical value and precision (right down to the definition of the exhibition materials for each section, including the *Boccalama* galley). The writing is unusually clear. It is an exhaustive project regarding the history of the Arsenal in its entirety, from its origins to the recent past. It is an enormous project in itself, involving consolidation works on a vast area (19,000 square meters of covered surface area, 28,000 square meters in total), and the complete refunctionalization of the southern part of the Arsenal. It also envisages the use of the water mirrors (a central resource of the Arsenal, sometimes forgotten), with the development of water itineraries. With an explicit logic of open access to the whole complex. It identifies a Library of the Sea, with archives and books about the civilization of the sea, from the State Archives and the Marciana National Library. The idea is then to entrust everything to an external agency to run the museum.

Among other things, the work of the Project Steering Committee (documented in the minutes of the Superintendency) describes a strong collegial effort, with the involvement of the President of the Republic himself (Ciampi), the Minister of Cultural Heritage and the Minister of Defense.

And yet nothing happened. The call for tenders, prepared by the Ministry of Defense (Ministero della Difesa, 2007), on June 6 was unsuccessful. It was not resubmitted and the project was definitively shelved, and it will never be resumed (indeed, there is no longer any question of it). Why? There is great embarrassment on the part of the actors at the time to answer this question, but one cannot evade the issue: the very possibility of recovering (in whole or in part) the important content of the project is at stake, even with the need to contextualize it within the current ownership situation. There are probably various aspects at play – one more conceptual, the other political.

As far as the conceptual aspect is concerned, there is an intrinsic weakness in the project – or at least in the design as developed up to that point, and which would have required further additions. The very fact that the project lacks a serious financial feasibility plan (incomplete investment costs and radical underestimation of operating costs) betrays this aspect. From this point of view, it is curious to follow the question of the link between the forthcoming National Museum of Naval History and the Naval Historical Museum of *San Biagio* that was mentioned at the start of this chapter. The explicit idea was to abandon the old museum because of its intrinsic weakness from a museological point of view.[4] This would have freed resources – monetary or at least figurative costs – in the presence of potential alternative uses of *San Biagio*, reducing the opportunity cost of its maintenance, in a logic of *value for money*. This would have been in accordance with Admiral Pagnottella's assessment of the intrinsic museological weaknesses of the old museum. In the end, however, any trace of this "saving" was lost, and in the final project, the two museums would have coexisted, with additional costs, and with the potential difficulties of explaining to visitors the relationship between the two.

But there is more, in the very way the tender was conceived, in relation to the degree of completeness of the project. In effect, up until then, it had been an incomplete plan: one that defined in great detail the needs for architectural intervention on the physical spaces, that had elaborated a museum concept of great depth, with details about exhibition materials, but without an evaluation of either the set-up costs or the likely operating costs. A call for bids that puts together the investment costs (€53 million – it is not clear how this was determined compared with the €57 million in the original document), does not define the set-up costs (there is a whisper of another €20–30 million) and underestimates the running costs can only go unanswered. In fact, it is not just a generic problem of imponderable risk, or in any case risk that has not been properly thought out. What is the overall investment, even if this is philanthropic? What is the income gap or the break-even conditions (for a museum of 19,000 square meters)? The problem is that two totally different kinds of intervention are being put together: the restoration of a considerable part of an immense complex (28,000 square meters) and the management of the outcome of this transformation (unfinished, even as an estimate). To think that a museum, no matter how excellent and well managed, could ever return an investment of this nature and size is simply naïve. Two different calls for proposals would have devoted attention to the search for two different actors: probably a philanthropist (institutional or personal) and a museum management organization.

Was this a technical error or a political issue in the preparation of this document? Here the search borders on gossip, on possible conflicts in this or that state apparatus. Certainly, the idea that someone wanted to kill the project – forever, without resuming it, which in itself supports the doubts – is more than plausible. Perhaps the case could lend itself to an interesting reconstruction of the institutional and personal interests of various stakeholders, which ultimately contribute, in one way or another, to a loss of historical opportunity (stakeholders are not always the best friends of an institution or a project). Moreover, the fact that some naïveté permeated the mind of the writer of the tender (and of those who approved it) is not unlikely, when asking (p. 9) the potential financier (of €53 million, plus set-up costs and management risks) to buy at a "price to be agreed" the PDF of the document we have described here!

Notes

1   Here a confusing issue emerges (not only in nominal terms). A museum close to the Arsenal already exists, and for a long time – the *Museo Storico Navale di Venezia* in *San Biagio*, owned by the Navy (www.marina.difesa.it/cosa-facciamo/per-la-cultura/musei/museostoricove/Pagine/default.aspx) but in recent years run by the Civic Museums Foundation, with the official English name as "Naval Historical Museum" (www.visitmuve.it/en/museums/naval-historical-museum/). The debate, though, was about a further museum of the Arsenal itself – see the play on words "*Arsenale e/è museo*", hard to translate, where there is a question whether the Arsenal should have a museum, though addressing the relation between the Arsenal "and" ("e") the museum; or the Arsenal "is" in itself ("è") considered as a museum.
2   "Venice has nothing but a small naval museum, which is moreover organized vertically, and one should wonder whether it is ever possible to organize a museum on the basis of verticality" (Pagnottella, 2002, p. 34).
3   The allocation of spaces is interesting, suggesting a museum where exhibitions and interactivity prevail (p. 48–49): 4,195 square meters for gallery-pathways and exhibition spaces; 2,719 for auditorium, multifunctional rooms, library and multimedia room, conference and projection rooms and spaces for training and meetings; 828 for restaurant, cafeteria and services; 1,230 for warehouses, documentation center, offices and laboratories (of which only 115 square meters is for warehouses!).
4   See Admiral Pagnottella's position already quoted in the second note, above. Moreover, the minutes of the Steering Committee refer to the "New location of the Naval Historical Museum" (February 25, 2002) and to the part "of the project that will provide for the redistribution of the material currently kept in the Naval Historical Museum in the new spaces that will be identified" (Technical Group Report, September 16, 2002) and to the "preliminary operations for the transfer of the Naval Historical Museum inside the Arsenal" (Zanelli's speech, minutes of the Committee of Arsenal Project Steering Committee, September 20, 2002).

Part IV

A research agenda

International perspectives on the Arsenal as tangible and intangible heritage

11 The Arsenal as intangible heritage

Between historical meanings and re-uses

Luca Zan

The Arsenal can be considered one of the most important industrial heritage sites in the world. By studying its internal complexity, we can focus the general debate on industrial heritage. A summary of the discussion in previous chapters will help address the issue of the intangible meanings of the site and questions about its valorization.

The debate on the Venice Arsenal: 40 years lacking a unitary view and actual results

Very many aspects of the Arsenal have been discussed over the past decades and represented in this book: unfortunately, they have lost consistency and incisiveness.

First is the historical value of the site itself (part I). It is one the of oldest shipyards in Europe, an engine for the development of Venice and centrally important to episodes in European history. Because it has been in use for 900 years, until very recently, the site permits an exceptionally extensive and detailed longitudinal analysis of many aspects, including decision-making processes. It has "outstanding universal value" (OUV) – to use UNESCO jargon – for studying management discourse between the 16th and 17th centuries especially, with interesting relevance in later periods. And it is relatively well preserved compared with other historical shipyards of its kind and size.

Second, such an unusual preservation situation – coupling authenticity and integrity – is the result of old and recent aspects of the site (see part II). It has been preserved over centuries (until recently) thanks to continuous manufacturing (and military) uses, despite huge transformations over time for technological and military updating (particularly in the 1800s). But unlike other similar areas, it has not been "bulldozed" or purposely destroyed for alternative uses, despite its central position. Problems did emerge with the

DOI: 10.4324/9781003200055-16

end of military production: a few decades after World War II, the decrease in intensive use and maintenance caused rapid decay and urgent need for immense preservation resources. But preservation legislation in Italy and local heritage norms did not allow "active destruction" for speculative uses of land, and most problems were due to lack of maintenance. Subsequently, a series of generous maintenance initiatives has occurred since the 1990s, particularly in the northern part of the Arsenal – although strangely, as we have seen, the various institutions involved have done very little to document their own efforts, and reconstructing "what" and "how much" has been done is difficult.

Third, a debate on possible re-uses of the Arsenal has occurred with unusual depth, both politically and socially (see part II). This has been quite a complex process: the two usual means of re-using space in industrial heritage (real estate and tourism) were barely applicable here. On the one hand, the enormity of the space as a whole and the huge dimensions of individual buildings (mostly hangars, covered pavilions, warehouses) make any alternative uses difficult and expensive. On the other hand, the general context of transformation throughout the city makes possible re-uses of the Arsenal for tourism more controversial compared with other places, because of systematically declining production activities, depopulation processes and associated over-tourism, with 30 million tourists per year, big ships and constant tourism-related water traffic all expected to grow worse in the future (Mancuso, 2009). The outcome of the process was an important achievement, the Master Plan, defining a composite destination of different areas, despite some problems: in particular, the lack of access for citizens and the removal of any idea of a museum was developed in parallel.

Fourth, the unstable role of the museum is also associated with the vicissitudes characterizing the development of major heritage projects (part III), as discussed earlier – for example, a silent competition between parallel projects ignoring each other, technical mistakes in the bidding process, yoking preservation investment, set-up costs and ongoing responsibilities in ill-defined operating conditions. If one basic element in the museum debate has been whether to open the Arsenal to citizens, the equally persistent problem of access has impeded any progress.

What should be underlined is the lack of consistency at different levels between various aspects covered in this book:

- There is no relationship between the museum debate and the re-use project emerging from the Master Plan (a missing link between parts II and III). Though partially due to weak feasibility assessments, the complete dismissal of any idea of a museum in the Master Plan sounds strange: some means of presenting the historical value of the site (part III) should be welcome.

- There is a lack of internal coordination between preservation and re-use (part II). Earlier restoration initiatives occurred largely in parallel, in silence, without coordination between them, and without an explicit long-term shared plan (the Master Plan was going to be approved concurrently, yet never executed). The risk of wasting money on renovations that then have no use is that they soon start to decay again (nothing is more dangerous for restored buildings than lack of use).
- There is a contradiction between the debate on the museum and the intangible meanings of the Arsenal's history (parts I and III). Little room has been left for understanding its management history and its "outstanding value" in the 16th and 17th centuries.

Valorizing the intangible meanings of the Arsenal

By analyzing the three museum projects (part II) and linking them to the historical meanings of the Arsenal (part I), it is easy to grasp the wealth of ideas, intuitions and suggestions that have occurred during years of debate. The specific lens of management focuses on two important problems that have not yet gained adequate attention, concerning the possibility of sustainably recovering and re-using this important site:

1 "Management in history" – a lack of attention to management (or *maneggio*, as in old archival documents) as a constitutive element of the history and modernity of the Arsenal.
2 "Managing history" – a lack of attention to evaluating the feasibility of each potential project, to evaluate what and how to manage in practice.

Management in history

There is a need to recover, research, interpret and later communicate not only the tangible part of the Arsenal but also its intangible meanings as heritage – a stratigraphy comprising centuries of knowledge, conversations and practices in management and accounting. Both tangible and intangible aspects call for "valorization", grasping the sense of what it means for an industrial heritage site to emerge as a productive organization requiring various intangible functions (skills, jobs, tasks, etc.), and in particular the even more intangible "meta-technology" of managing and organizing.

For administrative disciplines the Arsenal was truly an exceptional cradle of experimentation, with global importance, which should not be forgotten; this itself constitutes a specific "intangible" meaning of the Arsenal, its OUV. In this regard it is useful to remind that this extraordinary intangible significance of the Arsenal tends to be ignored, both within the management and accounting scholarly community and the heritage studies community.

Heritage scholars can be partially excused since they have less latitude to raise the issue of management than direct experts on the topic. But failure to meet these intellectual responsibilities among the international communities of management, accounting and business history scholars is simply unacceptable (see Chapter 1).

Given the international importance of management discourse at the Arsenal in the 16th and 17th centuries, there is a unique opportunity to present such an early example of modern management to citizens and visitors: a museum of management, or a section in the museum of the Arsenal focused on the *discorso del maneggio* and calling attention to the history of organizing. Such a gallery – whatever the space might be, not necessarily huge – would fill the serious gap highlighted by all the projects discussed earlier and would require few resources for adequate preparation.

The basic concept of such a gallery would concern the preservation, protection, research and interpretation of the intangible. While the protection of the tangible is already strengthened by legal standards, protecting and enhancing the intangible is much more difficult, sadly, and risks being forgotten. Additional research could be supported, exploring nine centuries of economic organization and the nearly unique transformations to the division and organization of labor. Certainly, interpretation would not be easy, because the continuously interacting transformation processes of both the tangible and the intangible. But there is an opportunity to curate innovative offerings, such as qualified guided tours of both tangible and intangible elements, which would generate value for citizens and tourists and create jobs (as with the experience of the Chicago Architecture Foundation). And, by increasing access to the site, open it to all citizens – which is one of the shared values is in the whole discussion.

Managing history

Any potential business model must be sustainable, for the whole site as well as for individual buildings, areas and re-uses (such as the museum). This is a huge challenge for (arts) management professionals and comprises a difficult but extremely interesting tension between the history of management and the actual management of historical sites including industrial heritage – in short, the "management of history".

The lack of attention to feasibility and sustainability issues from an institutional, organizational and financial standpoint is certainly not new in the cultural heritage sector. The failure of the overall discussion about the Arsenal museum is surely linked to such a lack of attention to sustainability: from the total absence of a single number in the *Arsenale e/è museo* project, to the simplistic budgeting of an under-defined project in the case of CSA, to the complete lack of forecasting for set-up and operating costs in the case of the Marina project. In turn, the Master Plan of the whole Arsenal site eliminates the idea of a museum, or even a minimal alternative like a visitors' center, without any explanation.

It is important to revive the whole idea of the Arsenal museum in order to break the deadlock of recent years. Crucially, the design of the museum must address the issue of feasibility. This will require resolving some of the main contradictions that have characterized the process, and learning from past experiences, by the following:

- Reconsidering spaces on a more modest scale, seeking tradeoffs in alternative potential uses (although still including a section on management).
- Being consistent about the relationship between the new museum and the existing Naval Historical Museum, perhaps even returning to the Navy's original proposal of merging the two into one, avoiding duplication and additional costs.
- Identifying collections and designing exhibitions that draw on the previous excellent work of the Navy's project; for example, acquiring and presenting the 14th-century *Boccalama* galley must not be further postponed.
- Taking a robust approach to preservation investment assumptions, set-up and running costs and income targets, as well as identifying adequate solutions for institutional design and governance structure.

Intangible and management studies: possible implications for the industrial heritage debate

The Arsenal case could address some important implications for the general debate on industrial heritage, opening the door to a more interdisciplinary vision of economic, business and management history.

The framework by Zan (2019b) could be extended. The relevance of the management perspective can recur cyclically within the perspective of a stratigraphy of organizing, at three levels: (1) its original meanings in historical time and evolution, that is, management as experienced in historical production; (2) the reconstruction, presentation and narratives of these meanings, that is, the degree of retention of the intangible significance in possible re-uses and (3), the managerial aspects of the valorization process, concerning the feasibility of management in re-use projects (Figure 11.1).

The relevance cycle of management in industrial heritage:
(a) management as experienced in historical production
(b) retaining the intangible significance in possible re-uses;
(c) the feasibility of management in re-use projects

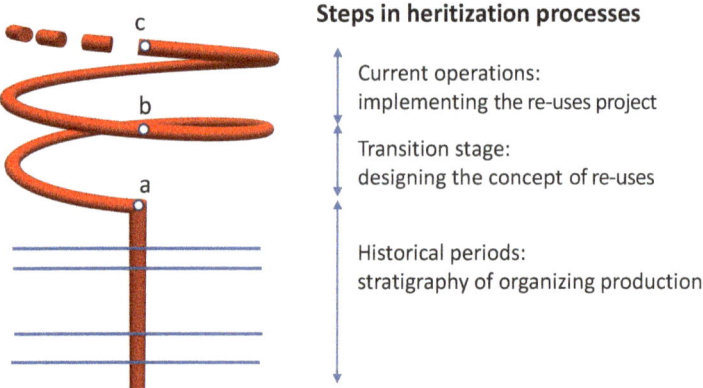

Figure 11.1 The management cycle and steps in heritization processes

In parallel, we can identify three conceptual steps in the process of heritization:

1   The reconstruction of the historical period – this involves the stratigraphy of a very long period of transformation (900 years for the Arsenal), where the main challenge is coupling the architectural transformation to its underlying (hidden, intangible) drivers: organizing processes as constitutive elements of industrial heritage (techniques, crafts, materials) and specifically managing the historical dynamics of getting thing done (including things now gone, as discussed in the Introduction). A new chronology could be developed, sensitive to organizing dynamics. The risk of forgetting the management or organizational factor – because it is invisible, hard to identify or requires archival "excavations" to reconstruct the stratigraphy of organizing – is a general phenomenon. Within this stratigraphy, a crucial element is the reconstruction of the final period of activity, when the process of halting (or significantly reducing) historical activities occurs (for the Arsenal, over the past 40–50 years).
2   The transition stage – this refers to the process of designing new re-uses and functions, starting from the end of production activities up to the

approval of a new concept for the site. In general, a crucial period could be the initial one, with the abandonment of part or the whole site, which will have important impacts on potential future re-uses (in the Arsenal case, this was after World War II, and it was not an extremely harmful period, thanks to the early experiment with *Biennale* and settlements in the north area). Next, assessment could be made of how much historical meaning should be preserved, and what kinds of narratives should survive in the new concept, along a continuum between musealizing the whole site and the radical removal of any sense of the industrial past. The case of the Arsenal is particularly intriguing here, because (a) instead of a single "site concept" there are several (the Master Plan of the site and the various museum projects) and (b) there are inconsistencies between them. In the former, almost nothing is preserved for interpretations and narratives; in the latter, much is retained, yet forgetting management and the organizing perspective.

3   Current operations of the new site – in principle, this will refer to the running of whatever site and new concept emerge from the design project at the end of the transition stage. Depending on the kind of transformation, this could be a "non-problem" (e.g., a building transformed into lofts and sold to private owners) or it could persist as a challenge to preserve the site with some attention to its historical meanings. In practice the process itself could be long and controversial, with the risk of in-progress transformations or slippage of the concept itself. Unfortunately, the case of Venice is an example of a process stuck in a limbo of worst rather than best practices, in which the Master Plan – with all its limits and contradictions – has not been executed following its design phase.

In the next two chapters we will develop further reflections for a comparative research agenda with other European historical shipyards, to support the processes of research and interpretation for similar sites by comparing trajectories of deindustrialization and re-use. The first analysis of structural and use differences (Chapter 12) helps to underline some specifics of the Venice Arsenal compared with other historical shipyards. This is followed by an attempt to structure a research agenda for studying the valorization practices of similar sites (Chapter 13).

12 The Venice Arsenal, singular but not unique

Materials for a survey of historic naval shipyards in Europe

Valentina Gambelli

The Venice Arsenal has not performed its original naval shipbuilding function for over a century. After a period of abandonment and subsequent degradation, a complex process of planning and redevelopment began, as described in previous pages (see also Bosio et al., 2017).[1] Now that this emergency phase of restorations has been completed, a unified vision for the complex is needed, further supported by renewed Listed Building status.

Outlining possible scenarios requires a broad viewpoint: the Venice Arsenal, while being an exceptional, probably unique example, is not the only surviving naval arsenal in Europe. Numerous questions emerge: how many navy arsenals exist today in Europe? Where are they located geographically? When were they built, and what are their dimensions? Do they still perform their original functions? Have their precincts remained closed, or are they publicly accessible? If they have been partially or entirely converted, what are they like now? Did accessibility and/or conversion contribute to any change in the relationship with the home city? Are there cases of coexistence between military and civilian functions that have accommodated urban permeability?

With the aim of answering these questions, in order to limit the investigation to those cases with an affinity with the Venetian one, certain criteria must be shared by all the arsenals considered:

- Original function – factories, or a suite of factories and spaces, originally built as maritime or riverine military naval arsenals. Merchant navy shipyards or commercial ports are therefore excluded. Conversely, arsenals founded as such, but which no longer have any military function, are included.
- Period – arsenals built from the Middle Ages onward, that is, contemporary with or later than the Venetian example. Arsenals from antiquity are therefore excluded.

DOI: 10.4324/9781003200055-17

• Integrity – the naval arsenals that are still in existence today. Lost or demolished arsenals although of considerable historical interest, are not relevant for contemporary conversion and transformation strategies.
• Location – the arsenals within the European continent, that is, those bordering the northern arc of the Mediterranean basin, the Atlantic Ocean, the Baltic Sea or the North Sea.

Moreover, even though they have been identified in the European survey map, arsenals that remain to this day exclusively military have been excluded for reasons of inaccessibility – for example, the Portuguese arsenal of Alfeite or the Greek arsenal in Salamis. The Spanish arsenal of La Carraca in the Gulf of Cádiz is also excluded; although it was built along with those of Cartagena and El Ferrol in the mid-18th century, on the instructions of the Bourbon King Philip V – and intended, like many of the Navy Command bases, to aid expansion of the Kingdom into the Mediterranean, the Atlantic and the Pacific – unlike the other two it is still fully operational and exclusively military. Perhaps it is not an accident that it is also the only one of the three that does not have a city on its flank.

So far, more than 40 shipyards meeting these criteria have been identified, spread evenly between the southern and northern shores of the European continent. Of these only five are riverine, 23 still perform functions related to the Navy and 19 still practice shipbuilding. Only seven are exclusively military. The remaining 16 also embrace other functions: museums, civilian shipbuilding and various urban functions (service, residential, etc.). Twenty-one shipyards have established a naval or maritime museum within their perimeter; among these, six are held exclusively by the military.

Return to port: reflections on the Venice Arsenal and the others

Comparing the Venice Arsenal with other sites invites reflection on its affinity, similarities and differences – in morphological-dimensional terms, regarding its current functions and considering conversion projects for the site.

No shipyard equates to that of Venice: individual themes or aspects suggest interesting analogies or thought-provoking strategies in other sites, but taken as a whole, the Venice Arsenal is certainly an exceptional case.

First, historical shipyards that have been divorced from their original functions are not all capable of being transformed into a museum. There are cases of medieval shipyards converted entirely to museums, such as those in Barcelona and Valencia. These are considerably smaller shipyards (the longest structure in the Barcelona shipyard is only half the length of the

Corderie in Venice); moreover, the integrity of the context has been lost due to urban infrastructure isolating them and distancing them from the water.

In cases where dimensions are greater and more complex, like the Venice Arsenal, a monofunctional solution proves unrealistic and reductive. Assuming a single large museum can guarantee the protection of a historical artifact oversimplifies a situation that instead needs an unfragmented strategic vision. The Venice Arsenal is very different, larger and more complex, than what remains of the shipyards of Genoa, Pisa, Amalfi, to mention only the maritime cities,[2] or of the 17th-century Palermo sheds,[3] or the medieval *Reales Atarazanas* (Royal Dockyard) in Seville[4] or the aforementioned yards in Barcelona and Valencia. All of these are undoubtedly unique artifacts transformed into exhibition spaces, but even a small selection of the *tese* within the Venice Arsenal would match any one of them for size.

The Venice Arsenal also, unlike the others of medieval origin, has preserved much of its historical heritage, because continuous production activity through the centuries has entailed constant maintenance of the factory buildings.

Generally, a change from military to civilian use has coincided with transfer of ownership and subsequently with the contraction of the military precinct in favor of the expansion of the neighboring city. This has sometimes led to decisions based on the physical enclosure of the military holding, rather than on the ideal extent of the museum.

In the case of Cartagena, for example, the southeastern area of the shipyard became public and the wall that originally separated it from the city was demolished. Simultaneously, former prisons were converted into a polytechnic, foundries were restored as a naval museum and an elongated square that lies between the two was reconfigured. But, whereas on the southern front, this opened a welcome view toward the sea, on the opposite side, the demolition of the diagonal was controversial. The significant transformation of this Spanish shipyard should also be contextualized in the larger project of redefining the entirety of the city's waterfront,[5] for which it provides a natural boundary to the west.

Concerning the redevelopment of open spaces and pathways, it is worth mentioning the interesting example of a competition launched in 2015 for the historical portion of Her Majesty's Naval Base Portsmouth, UK, won by Latz + Partner, persuasively addressing issues common to all shipyards where historic buildings have been restored and opened to the public: accessibility, hierarchies of routes, stop locations, lighting and underground services in general.

The theme of readability of open spaces is sometimes closely linked with the more general visions of the city that hosts them. For example, in Pisa, the Arsenals of the Republic have been involved since 2015 in PIUSS, a project of the Municipality of Pisa which picks up the baton from the Park of the

Citadel, an uncompleted 1957 project by Giovanni Michelucci ("City Walls project: restoration and enhancement of the fortified system").

In Lisbon, PROAP (João Nunes, 2009–2013), a project to recover the open spaces of Lisbon's historical shipyard, should be read as a single phase of the redevelopment of the entire north bank of the Rio Tejo. Naval shipbuilding along the *Ribeira das Naus* was never confined in a circumscribed space but instead was dotted along the shore and connected by water; this is why the Rope Works, commissioned by Pombal in 1771 after the earthquake that destroyed the city, are four kilometers west of the shipyard.[6]

The Venice Arsenal is also unique because it is inaccessible by car and can be reached only by the same means available centuries ago: on foot or by boat. Though banal, this has probably been a significant factor in its preservation. As we know, there has been no shortage of attempts on territory to enable more direct (and faster) links with the mainland infrastructure, but the preservation of the lagoon and the historic city has always prevailed. Roads and railway lines remain more fantastical than likely.

There are cases, however, in which the transformations are the result of a pragmatic resident-centric process, linked to the design of the infrastructure network: the city grows and is planned, existing spaces are colonized, history marches on. This is the case of the Amsterdam shipyard where in one of the recent radical modifications a parcel of land belonging to the Kattenburg Marina was hived off to create space for the entrance to the IJtunnel. (The Science Museum NEMO was then built on top of the entrance by Renzo Piano Building Workshop.) The IJtunnel, excavated under Lake IJ in 1967–1971, links the historic center of Amsterdam and the Amsterdam-Noord expansion. One wonders whether there could have been alternative approaches to urban development better respecting the traces of the past – whether, somewhere between reverential freezing in time and complete erasure of memory, we can practice other methods capable of embracing complexity as an asset, not an obstacle.

When compared with other shipyards, the density of the built environment in the Venetian example is surprising: in Copenhagen, for example, where in the 17th century a system of islands was developed specifically for the new shipyard covering a third of the fortified city, the building density envisaged was relatively modest and the empty spaces for open water sheds were very extensive. This is the opposite of what happened in Venice, where the number of covered sheds has no equal. The Danish settlement model, in the centuries following initial construction, favored a progressive urbanization that did not (always) require demolition of pre-existing buildings. Some notable pieces of architecture, such as the Ordinance Hall, or the surviving twin of a pair of *masterkranen* (masting crane) at Holmen Naval Base on the island of Nyholm or the array of wooden gunboat sheds, today provide a counterpoint to the process of densification and residential transformation.

Dimensions and use situations: a first comparison

Figures 12.1, 12.2 and 12.3 show individual European shipyards on satellite images in comparison to the Venice Arsenal, an aerial photo of which is shown at the head of each table at the same scale. The shipyards shown here are, in order of foundation, Amalfi (11th century), Venice (first half of the 13th century), Pisa (Republican Arsenal early 13th century and Medici Arsenal first half of the 16th century), Seville (1252), Genoa (1283), Hvar (1292–1331), Istanbul (1455), Portsmouth (1490), Lisbon (1498), Palermo (1621), Amsterdam (1656), Marseille (1669), Copenhagen (1690–1780), Pola (1856–1858), Augusta (1900–1934), Kiel (1957) and Wilhelmshaven (1957). The red perimeter line, where dashed, indicates either original areas that have suffered material loss or, in most cases, sectors of the shipyard that are no longer under military control but which survive, physically and/or in outline.

This apparently simple operation emphasizes the dimensions of the Venice Arsenal compared with the others, also allowing a glimpse of the location and surrounding urban contexts of individual arsenals. What is immediately noticeable is that the extant factory buildings alone cover a considerably larger area, when compared with the Arsenal's peers. This is due to a series of successive enlargements over the centuries, whereas the other medieval arsenals discontinued their activities much sooner.

These comparisons also prompt a more specific investigation into the temporal-dimensional aspect: the comparison of shipyards in their various historical phases, focusing on the evolution (or involution) that has preserved their physical integrity over time.

While a diachronic, or evolutionary, view of European shipyards can highlight how certain transformative historical events are surprisingly relevant today, a synchronic view looking at specific points in time can capture any civic responses to the changing needs of the shipbuilding industry.

It would lend clarity, for example, to the outcome of the events that simultaneously befell the Venice Arsenal, La Spezia and Taranto, when, in the aftermath of the unification of Italy, there was a program of reorganization of the newly born Royal Navy, attempting to control the Tyrrhenian, the Adriatic and the Ionian. In Venice the existing shipyard was expanded between 1866 and 1885, while those of La Spezia and Taranto were built *ex novo*, in 1862–1869 and 1882–1889 respectively. Allowing for specific differences, each site shared modern requirements for the construction of battleships:

> [The] structures with which a modern shipyard must necessarily be equipped mainly constituted volumes of water of a size and depth appropriate to the construction of ships larger than 100m in length; quays permitting easy arming, loading and unloading operations; slipways to accommodate open-air construction of large ships; dry sheds

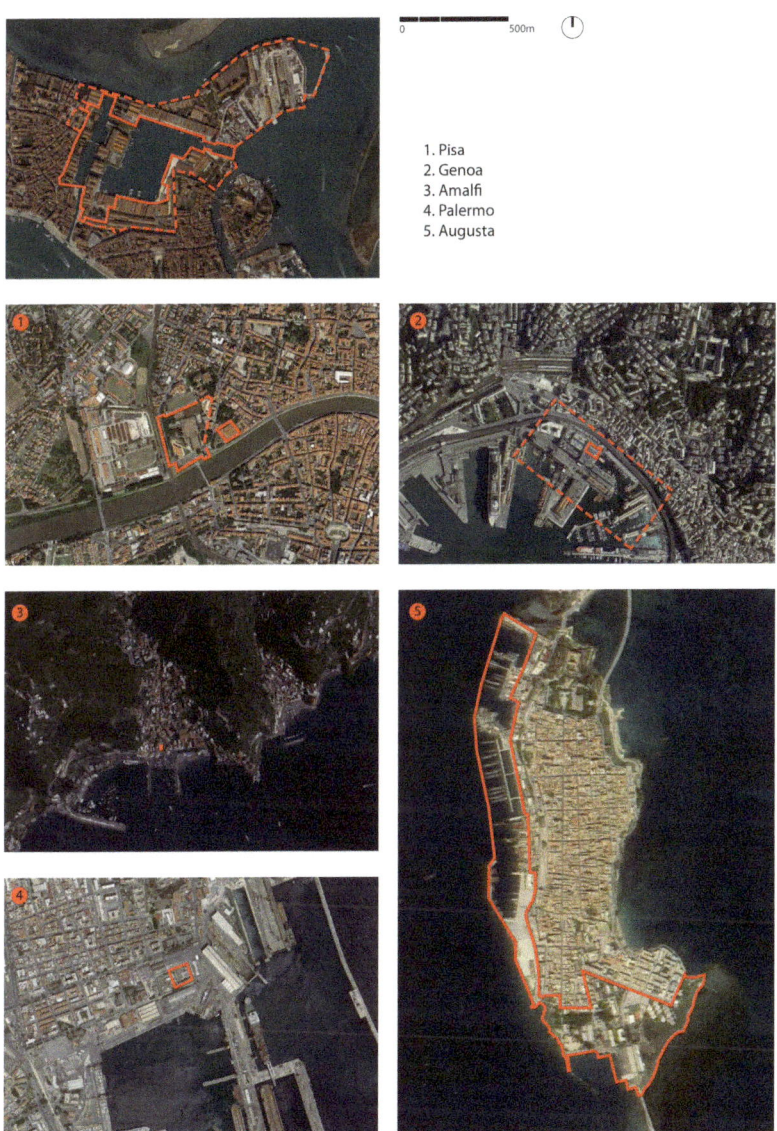

1. Pisa
2. Genoa
3. Amalfi
4. Palermo
5. Augusta

Figure 12.1 European historical shipyards 1

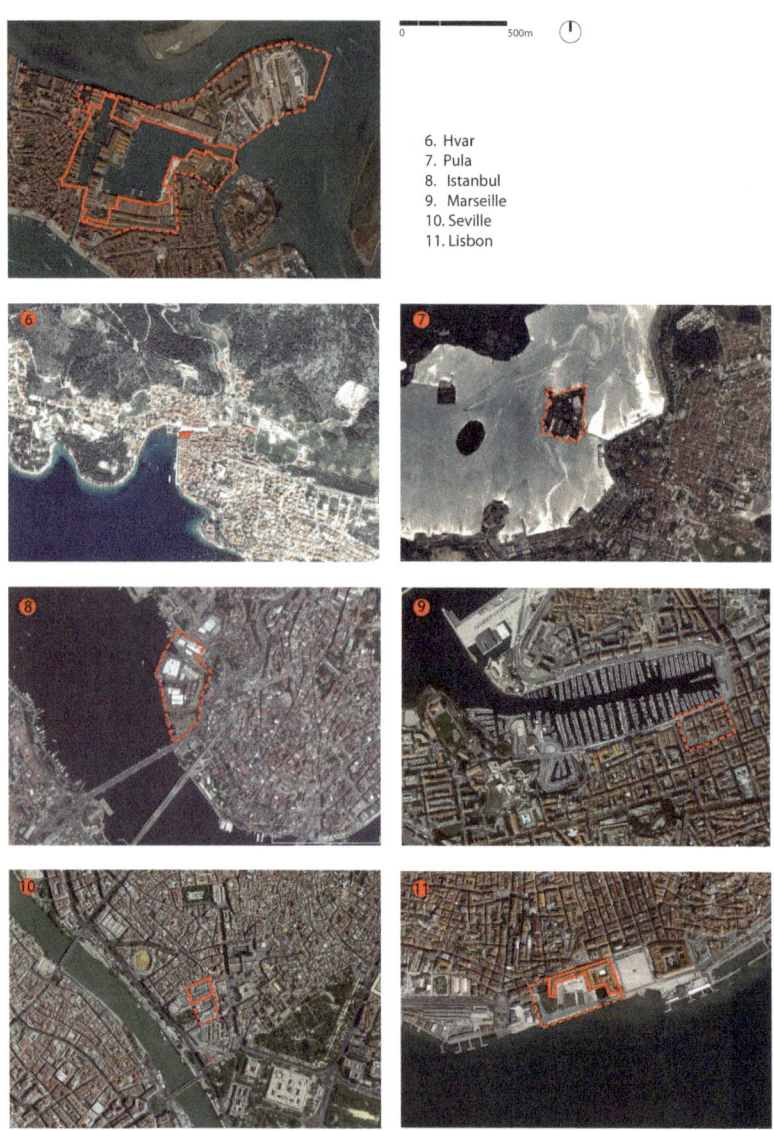

Figure 12.2 European historical shipyards 2

0   1km

12. Portsmouth
13. Kiel
14. Wilhelmshaven
15. Amsterdam
16. Copenhagen

Figure 12.3 European historical shipyards 3

for maintenance and repair of hulls out of the water; lifting machinery (cranes and overhead travelling cranes) for handling weapons, armour plating, and heavy machinery; and specialized workshops for manufacturing and precision engineering.

(De Maestri et al., 2018)

In conclusion, this first excursion compares the Venice Arsenal with shipyards around Europe, exploring both affinity and differences, in terms of age, morphological conditions, historic artifacts preserved and size. As a result, we can identify strategies adopted and adoptable for managing the inevitable transformative process affecting them.

Notes

1 As early as 1986, the Venice Arsenal was protected pursuant to law n. 1089 of 1939, but by 2010, the Superintendent of Heritage for Venice and the Lagoon had declared the Arsenal Complex to be *of cultural interest*, thereby extending the protection to all aspects which bear witness to the cultural value of the complex: not just buildings but also open spaces and machinery.
2 Galata Museum, Genoa: inaugurated in 2004, designed by G.V. Consuegra. Pisan Republican Arsenals: the surviving shipbuilding sheds were restored as a space for exhibitions and conferences in 2011–2015. A Museum of Ships is planned within the Medici arsenals. Both are Municipal initiatives. Amalfi Arsenal: what remains has housed the Museum of the Compass, a space for temporary exhibitions, (discontinuously) since 2010.
3 The arsenal of the Royal Docks, managed since 2013 by the *Soprintendenza del Mare*, now hosts associations, activities and exhibitions related to maritime heritage.
4 El Arsenal de la Real Marina: decommissioned 1593; re-purposed 1641 as Hospital of Mercy; now a museum of the same name. In 1993, the Ministry of Culture began minimal restoration of the remaining sheds for cultural and exhibition uses, and in 2009 announced a design competition to redevelop the entire building. The winning entry, by a group headed by G.V. Consuegra, has not yet been realized.
5 On the seafront at Cartagena several architectural projects have been developed: the National Museum of Maritime Archaeology, by Consuegra, 2009, the *El Batel* Congress Centre, by SelgasCano, 2002–2011, and the extension to the cruise ship terminal by Lejarraga, 2012.
6 The Rope Works in Lisbon currently houses the Municipality's temporary exhibitions and, since 2007, the *Trienal de Arquitectura de Lisboa*: playing exactly the same role as the Venetian *Corderie*.

13 The heritization of historic naval shipyards in Europe

Paolo Ferri, Pegram Harrison and Luca Zan

Chapter 11 provided a general framework for interpreting processes of heritization in general, and a first application to the complex processes of the Venice Arsenal, from its earliest history to the (unfinished and somehow contradictory) re-uses proposed over the past decades for deindustrialization and partial demilitarization. Chapter 12 provided a preliminary overview of historic naval shipyards (HNS) in Europe, with important observations on patterns of change, mainly from the point of view of the architectural transformations affecting them. Following the same path, this chapter provides the basis of a comparative research agenda for HNS, in an attempt at extending the framework to industrial heritage in general. First, we make the case for comparative research with a focus on managing and organizing. Next, we define the focus of the research and discuss sample selection, data collection and analysis. Finally, we outline the expected impacts of this research agenda.

Historic naval shipyards and their relevance

Historic shipyards have had enormous impacts on the economic and social evolution of vast areas, particularly in Europe, a region whose relative prominence in global affairs derives in large measure from maritime institutions and infrastructure (Braudel, 1949). Research on their historical origins alone is worthwhile.

But HNS sites are more than mere metaphors for dynamically changing environments; they pose crucial issues for preservation, protection and use, with significant socioeconomic impacts. Their changing uses – from production to other activities (e.g., exhibitions, shops, housing) – have detached them from their original meanings and imbued them with new ontologies that somehow co-habit with the old ones, leading to tensions between preservation and innovation. Practically, this has also brought managerial challenges far subtler than generating income and managing visitor flows: sites

DOI: 10.4324/9781003200055-18

must trade-off the different and often conflicting priorities of preservation, productivity, housing, visitor enjoyment, community engagement, and so forth. Moreover, these sites are at once exceptionally old, newly construed as heritage (both tangible and intangible), and conflictual given the variety of uses and values attached to them. Also, changing uses have caused significant shifts in their physical features, organizational purposes, and social engagements. Thus, different narratives can be told at different times and at different levels about buildings, organizations and communities.

Consequently, research related to HNS is highly fragmented across disciplines and locations, and studies tend to be national and site-specific with limited transferable relevance. But it is possible to take a different view that focuses on historic shipyards as a crucial element in understanding the history of European societies through time, first as places of innovation and growth, and then as sites of deindustrialization and (sometimes) regeneration processes. This variety of processes across time is particularly interesting, as narratives about the past are constructed through patterns in heritization.

Patterns in heritization

While reconstructing the architectural transformation of such sites is a preliminary step that provides some basis for investigating the more complex socioeconomic process of transformation, here we address the variety of processes in heritization itself, looking at preservation of ancient artifacts and ongoing re-uses of the site due to processes of deindustrialization and demilitarization. All of this implies an issue of degree: preservation, re-uses and deindustrialization can refer to the whole site or just a part of it.

Our perspective implies some conditions for the inclusion or exclusion of sites in the research agenda (following the suggestions by Gambelli in Chapter 12):

- The reduction of historical uses – sites are relevant for our research only if they have experienced to some extent a process of demilitarization or deindustrialization; sites are excluded if they are still used exclusively for original manufacturing and/or military activities.
- At least partial integrity – sites are excluded that have not been substantially destroyed, because they present no tensions between re-use and heritization.
- A question of re-use – sites are included where possible re-uses and the preservation of elements of previous historical uses are being debated (particularly narratives about their intangible meanings).

The three layer-model presented in Chapter 11 provides a framework to investigate each HNS case study and to inform comparison for identifying common and distinctive elements at each site. The framework structures the process of heritization (old uses, no uses, re-uses and current uses) and focuses on three areas that can be operationalized as follows: (1) the reconstruction of historical uses, (2) the transitional stage, (3) the current operation of the site as heritage.

1 The *reconstruction of historical uses* includes the history of buildings *and* organizational practices. For buildings and spaces, changes in physical structures can be understood by mapping structures across space and time to reconstruct patterns of changing historical uses (yet within the manufacturing/military phase, before deindustrialization); relevant sources for such research would comprise papers, books, maps, projects and other archival records. For organizational practices, their history can be grasped by comparing varieties of institutional structures and their long administrative evolution, plus context-specific and path-dependent transformation processes in historical sites of industrial production, to reconstruct the evolution of organizing over time. In addition to papers and books, archival records and oral history (when possible) would represent valuable sources to trace such changes and turning points in organizational history. For both perspectives, architectural and organizational, several crucial periods would require specific attention: in particular the "last layer" of the historical period, a process in which a site ceases its historical "old uses"; a possible following phase of "no use", a sort of limbo where old uses no longer happen; and new uses not yet considered, a critical period that could easily end with the destruction of the site as a whole.

2 The *transitional stage* is the period in which a discussion of possible re-uses emerges and develops. It is in principle an intermediate period, between the ending of the previous, historical period, and the new life of the site as an industrial heritage entity. The issue of defining "re-uses" of the site is crucial, with several important implications. Research must address issues of protecting and representing a plurality of narratives converging on historic shipyards and consider how different stakeholders perceive and represent the sites with multiple sets of meanings. The research questions at this stage would ask: what are the established protection policies and practices for adaptive re-use across unmovable, movable and intangible heritage in historic shipyards? What kinds of structural and spatial allocations are envisioned? What degree and what kind of musealization serve what narratives? What are the technological paradigms for re/presentation? Research sources for evaluating patterns of protection for each site would include legal frameworks, regulations, and explicit policy projects. Interviews with curators would help to understand actual practices. And a mix of direct

observation, document analysis and interviews would elucidate the ways in which meanings of each site are constructed and represented. What is particularly intriguing from the perspective of organizing is how the intangible aspect is retained, preserved and interpreted to potential users.

3 The *current operation of the site as heritage* follows the transitional phase, once "new uses" have been established. New research questions are worth asking here. Some concern sustainability: is any public discussion of sustainability is still open? In general, a complete process of privatization (e.g., if sites have been converted into residential properties and sold to private owners) implies the end of the whole discussion, with some minimum level of respect of protection laws in terms of restrictions on demolition. Conversely, if some public involvement survives, then other issues are relevant: institutional design (How do different entities run different spaces?); governance structures (What are the levels and mechanisms of authority?); financial resources (What investments and operations support each site's heritage components?); human resources (What skills and competencies relate to heritage?); and day-to-day organizational practices relating to heritage (What cost controls, budgeting, leadership, organizing labor and marketing practices occur?). Statutes, contracts, internal documents, interviews with relevant actors and observation would be the ideal data sources for this area of the research.

A truly transdisciplinary research agenda would look through all of these lenses together, identifying patterns of challenge and response over time and at each layer.

Expected impact

The proposed comparative research on HNS heritization processes has the potential to make significant impacts in conceptual and instrumental ways, as well as in capacity-building and connectivity. In this final section we explain how, and who will benefit.

The research aims to achieve conceptual impacts by engaging in knowledge exchange activities with a wide community of stakeholders by supporting a range of dialogues with policy-makers, businesses, community groups and the general public, and by promoting the emergence of distinct, sometimes conflicting, narratives around HNS. A systematic international comparison of sites, looking at similarities and differences – not merely in final outcomes but also in the reconstruction of old, unchanged, new and current uses – will facilitate broadly relevant learning. Also, the research will promote the sites as European-level heritage, enhancing awareness and exploring new ways of engaging citizens.

Expected instrumental impacts would include changes in HNS-related policy and professional practices. Evidence produced by cross-country comparison of established policies for adaptive re-use of unmovable, movable and intangible heritage would change and strengthen policy frameworks and professional practices for protecting and representing HNS, especially its intangible heritage. The research would also establish best practices in HNS management, leading to improvements in decision-making, strategies for investment and operating revenues, cost controls, institutional and governance solutions and human resource management.

Potential capacity-building impacts would highlight the optimal professional skills needed for collecting, protecting, representing and managing historic shipyards. This will benefit primarily site managers and heritage specialists and create opportunities in higher education to redefine curricular offerings.

Last, connectivity impacts would foster enduring relationship among stakeholders, bringing together policy-makers, businesses and community groups within each site, and establishing links between sites. This would improve knowledge exchange and the diffusion of best practices – for the lasting benefit of these enormously important sites and all who benefit from them.

The previous chapters of this book prove how much all of this is needed.

Conclusions

A policy agenda to not forget

Luca Zan

In addition to underlining the importance of the Arsenal in terms of management history, this book reconstructs debates on recovering the Arsenal, with efforts in preservation and designing potential re-uses for the future. A milestone was the production of a Master Plan, unanimously approved by the City Council in 2001, and then updated to 2014. In parallel, a tangled discussion about a new Museum of the Arsenal took place, with various projects that were uncoordinated and competed with each other. Among these, the idea of a National Museum of Naval History mobilized a considerable amount of energy and became a project of great scope that inconvenienced a couple of ministers and the President of the Republic of the time himself.

What happened then? Nothing. On both levels – the overall recovery of the Arsenal and the museum – silence fell, and there was a deep and total removal of the projects. Both were forgotten. For the museum, it was already dead in the 2014 version of the Master Plan, which simply removed it. The more complicated story is that of the removal of the Master Plan itself. In addition to intrinsic difficulties of the challenges of transforming a site of this nature, institutional fragmentation and managerial collapse were among major causes – made worse by the Law 221/2012 that transferred the ownership to the Municipality in peculiar ways, with the winding up of SAV (see part II).

Soon after the transfer, local government fell into a deep crisis and the mayor resigned, which led to a period of extraordinary administration, the so-called *commissariamento*, where a representative of the Ministry of Internal Affairs was in charge of running the city for a period, up to a new election. Then, a different political coalition won and ran the city. The Arsenal was "not a priority in our political program", stated the representative of the mayor in a conference at the Ateneo Veneto in April 2017. In parallel, the Arsenal Office of the Municipality, which had replaced the SAV it wound up, was eventually closed down. The Master Plan was never mentioned again, while a new project of running a 10-day Venice Boat Show was launched in recent years.

In parallel, Venetian civil society was experiencing a variety of initiatives. For what matters here, the association *Forum Futuro Arsenale* (2016) developed an alternative project for re-using the Arsenal, which is also available in English on its website. At the very least, this was a set of interesting suggestions that could have been taken into account in some way, although they lacked an adequate feasibility study in terms of financial costs – but it was ignored.

In the absence of support for the overall Master Plan, even the idea of the innovation hub of the Master Plan (see the green area, #1, Figure 6.3) slowly started winding down. Some of the institutions that were planning to move there changed their plans; part of the buildings *(Tese delle Nappe* and *San Cristoforo)* have been transformed into spaces to rent for commercial use by the Municipality (weddings, parties, company events).

A more recent change also affects the northern part of the Arsenal: the decision to move to the mainland the activity of maintenance of the MOSE bulkhead, which will free the whole northwest area (see the yellow area, #3, Figure 6.3) for industrial activities to be identified and developed. What remains in question is the future of CVN, with possible implications also for the uses (and its right to use spaces) that emerged in the period following the transfer of ownership by the Law 221/2012.

There are undoubtedly limits and contradictions in the overall planning process of the past decades, both at a general level and at the museum level, and above all in terms of (lack of) access for citizens. Nonetheless, the total abandonment of these projects and plans risks scattering intellectual capital (and shared achievements) to the wind. Decision-making processes have to start from scratch every time, which contributes to the chronic situation of doing nothing. At present, neither journalists nor politicians seem to remember a debate that was rich and interesting and was then buried at the level of city quarrels.

Is it possible that the most important industrial heritage site in the world is still so neglected? That so much has been done on preservation, but there have been no decisions on uses and re-uses – or that the Master Plan has never been executed, with the risk that buildings that have been renovated with great skill and attention will soon decay due to non-use (e.g., the Tower of *Porta Nuova*)? That a galley dating back to 1300 is still lying under the lagoon without any resources or space being found in 20 years to recover it and make it accessible? That the submarine *Dandolo* – twin of the *Toti* submarine at the Museum of Science and Technology in Milan, with a daily queue of visitors providing not insubstantial income – remains closed and not accessible, in an area itself not accessible to the public and to citizens? That to organize a visit to the Arsenal, one is forced to beg hospitality from five different organizations, instead of being a right of the citizen?

A "policy agenda" is needed, where technical details are the less complicated issues – despite the challenges of a not trivial updating. The problem of governance is still crucial, in the absence of a strong political commitment in which main institutions can return to "collaborate" – especially the *Biennale* and the Navy, while CVN is becoming a sort of ghost entity. The political willingness, on the part of the main owner, the Municipality, is the starting point to set up this policy agenda.

A final comment: the book addresses huge deficiencies in management and leadership across 40 years of talks about the Venice Arsenal – at the level on preservation, defining re-uses, the museum issue and the overall policy agenda – at a site that has been one of the cradles (or is at least an early example) of modern management. Quite a paradox. Baldissera Drachio probably turns in his grave.

Epilogue

Luca Zan

Just as the book is going to print, important news involving the Arsenal appeared in the press. On December 23, 2021, the Venice City Council approved an agreement with the Ministry of Defense and the Ministry of Culture for a project of valorization of the Arsenal (Comune di Venezia, 2021). Basically, the *sine die* area will be carved up: one part will go back to the Navy (against what was defined by Law 221/2012); the rest, eventually ceded by the Navy, will be used by the *Biennale* and a project by the Ministry of Culture for a center of excellence on contemporary art. The Municipality of Venice will no longer be involved in this area, and as a counterpart, it would have access to the water basin two weeks a year to organize events, plus the possibility of using the *Rio dell'Arsenale* for public motorboat passage. There is no news about re-uses of the northern part, and nothing new in terms of free access to the site by citizens.

All of this simply forgets the overall debate of the past 40 years, described in this book, basically ignoring all decisions made to protect the area (including the stop to motorboat passage, and the destination areas defined by the 2001 Master Plan), proposals and debates (see the museum's projects or the proposal by the civil society and grassroots organizations). On top of this, a new "strategic" project around the *Biennale* is imposed, with serious issues of transparency if not democracy. After more than eight months from the initial communication by the Ministry of Culture (MIC, 2021), nothing is yet available about the new project (despite a commitment of €170 million), or its sustainability, including running costs once the design phase is over (as well as extraordinary EU funding).

In any case, the preservation of the Arsenal and the maritime identity of the site are further threatened by these initiatives, and barriers to access for citizens are even increased. Indeed, it is a good example of worst practices in heritage management.

References

AA.VV. (1986). *Venezia e la difesa del Levante: da Lepanto a Candia, 1570–1670*. Venezia: Arsenale Editrice.

Agnoletto, M., Pasqualetto, C. (2016). *Museo N.A.VE. – Prospettive sospese*. Tesi di laurea magistrale, relatore Stefano Rocchetto, IUAV, Venezia.

Ansoff, H. (1984). *Implanting strategic management*. Englewood Cliffs, NJ: Prentice-Hall.

Ari, B., Zan, L. (2021). *Shipbuilding and early forms of modern management: Insights from the Ottoman reconstruction of the fleet after the Lepanto defeat in 1571*. Working Paper University of Bologna.

Balzani, R. (2004). *Per le antichità e le belle arti: La legge n. 364 del 20 giugno 1909 e l'Italia giolittiana*. Il Mulino, Bologna.

Bellavitis, G. (1983). *L'Arsenale di Venezia: Storia di una grande struttura urbana*. Venezia: Cicero Editore.

Biadene, S. (a cura di) (1990). *Carte da navigar – Portolani e carte nautiche del museo correr 1318–1732*. Venezia: Marsilio.

Bosio, M., Fornasiero, T., Gambelli, V. (a cura di) (2017). *Arsenale di Venezia: Progetti e destino*. Conegliano: Incipit Editore.

Braudel, F. (1949). *La Méditerranée et le Monde Méditerranéen a l'époque de Philippe II*. Paris, A. Colin.

Buratti, B. (2019) (a cura di). *Francesco Morosini. 1619-1694. L'uomo, il doge, il condottiero*. Roma: Poligrafico e Zecca dello Stato.

Cabianca, A., Pellizzari, R. (2016). *L.A.MA. Laboratorio di archeologia marina*. Tesi di Laurea magistrale, IUAV, Venezia.

Canal, E. (1978). Localizzazione nella laguna Veneta dell'isola di San Marco in Bocca Lama e rilevamento di fondazioni di antichi edifici. *Archeologia Veneta*, I, pp. 167–174.

Canal, E. (2013). *Archeologia della laguna di Venezia*. Venezia: Cierre Edizioni.

Caniato, G. (2002). Il Progetto Arsenale della Civiltà dell'acqua. *Quaderni Insula*, 11, pp. 27–30.

Caniato, G., Fumagalli, E. (a cura di) (2002). Arsenale e/è Museo: Due modi per un uso unitario. *Quaderni Insula*, 11. www.insula.it/index.php/quaderni/106-arsenale-ee-museo-11-2002 (Accessed 18.12.2021).

Casoni, G. (1829). *Guida per l'Arsenale di Venezia*. Venezia, Tipografia di Giuseppe Antonelli.

Castelli, E. (2002). Presentazione. *Quaderni Insula*, 11, pp. 3–4.

Chandler, A.D. (1986). Gli Stati Uniti: L'evoluzione dell'impresa. In Payne, P.L., Kocka, J., Yamamura, K., Chandler, A.D. (eds), *Evoluzione della Grande Impresa e management* Torino: Einaudi (ed orig Cambridge 1978), pp. 5–88.

Chirivi, R. (1976). *L'Arsenale di Venezia: Storie e obiettivi di un piano*. Venezia: Marsilio Editori.

Città di Venezia (2014). *Documento Direttore per l'Arsenale di Venezia – 2014*. Direzione Patrimonio e Casa. http://arsenale.comune.venezia.it/?page_id=319.

Città di Venezia (2015). *Documento Direttore per l'Arsenale di Venezia*. Venezia: Marsilio.

Città di Venezia (2017). *Progetto di governance territoriale del turismo a Venezia*. http://live.comune.venezia.it/it/2017/07/progetto-di-governance-territoriale-del-turismo-venezia-scheda-di-sintesi (Accessed 02.01.2018).

Comune di Venezia (2021). *Via libera dalla Giunta al Protocollo d'intesa con i Ministeri di Difesa e Cultura per il progetto di valorizzazione dell'Arsenale di Venezia*. https://live.comune.venezia.it/it/2021/12/libera-dalla-giunta-al-protocollo-dintesa-con-i-ministeri-di-difesa-e-cultura-il-progetto-di (Accessed 18.01.2022).

Concina, E. (1984). *L'Arsenale della Repubblica di Venezia*. Milano: Electa.

Concina, E. (2004). L'Arsenale di Venezia: spazi, fabbriche, funzioni (1220–1813). In Dina, A. (ed), *La rinascita dell'Arsenale: la fabbrica che si trasforma*. Venezia: Marsilio, pp. 68–79.

Crumlin-Pedersen, O. (2002). La rivitalizzazione del patrimonio marittimo danese: L'esperienza di Roskilde. *Quaderni Insula*, 11, pp. 15–20.

CSA & CNR, (2006). *Studio di prefattibilità Museo della Cultura e della Civiltà del Mare – Arsenale di Venezia*. Versione 3.0. Venezia: Marsilio.

CVN, Consorzio Venezia Nuova (2002). *La galea ritrovata: Origine delle cose di Venezia*. Venezia: Marsilio.

Davis, R. (1991). *Shipbuilders of the Venetian Arsenal: Workers and workplace in the pre-industrial city*. Baltimore: Johns Hopkins Press.

De Maestri, S., Menichelli, C., Monte, A. (2018). La storia parallela degli arsenali di Venezia, La Spezia e Taranto, dall'Unità d'Italia a oggi. In Fontana, G. (a cura di), *Stati Generali del Patrimonio Industriale*. Venezia: Marsilio.

Fletcher, C.A., Spencer, T. (eds) (2005) *Flooding and environmental challenges for Venice and its Lagoon: State of knowledge*. Cambridge: Cambridge University Press.

Forum Futuro Arsenale (2016). *The Venetian Arsenal and the city*. https://farovenezia.files.wordpress.com/2016/10/progetto_arsenale_en.pdf (Accessed 10.01.2022).

Gambelli, V. (2017). Trasformazioni recenti: Schedatura e disegni dei progetti realizzati. In Bosio, M., Fornasiero, T., Gambelli, V. (a cura di), *Arsenale di Venezia: Progetti e destino*. Conegliano: Incipit Editore, pp. 62–175.

Gelderblom, O., Trivellato, F. (2019). The business history of the preindustrial world: Towards a comparative historical analysis. *Business History*, 61(2), pp. 225–259.

George, C.G. (1972). *The history of management thought*. London: Prentice Hall.

ISMM, Istituto di Studi Militari Marittimi (2007). *Progetto arsenale: Museo nazionale di storia navale. Progetto di fattibilità*. Venezia: ISMM.

Johnson, H., Kaplan, R. (1987). *Relevance lost: The rise and fall of management accounting*. Boston, MA: Harvard Business School Press.

Lane, F. (1934). *Venetian ships and shipbuilders of the renaissance*. Baltimore: Johns Hopkins Press.

Lombardi, G., Paternò, R. (1992). *Riqualificazione dell'Arsenale il quadro urbanistico*. www.fondazionevenezia2000.org › 2007/09/17 › riqualificazione-dell-ars... (Accessed 10.01.2022).

Mancuso, F. (2009). *Venezia è una città*. Venezia: Corte del Fondego.

March, J.G. (1988). *Decisions and organisations*. Oxford: Basil Blackwell.

Marsala, H. (2014). Venezia, tre giorni di storia e di cultura con l'Arsenale Aperto alla Città. *Artribune*.

MIC, Ministero dei Beni Culturali (2021). *Cultura – "Next generation Eu: Recovery & resilience plan"*. https://cultura.gov.it/recovery (Accessed 18.01.2022).

Ministero della Difesa – Direzione Generale dei Lavori e del Demanio. *Avviso indicativo di finanza di progetto (artt.152 e seguenti del D.Lgs. n.163/2006 e s. m. i.)*. www.difesa.it/SGD-DNA/Staff/DT/GENIODIFE/Bandi/Pagine/VENEZIA_1530. aspx (Accessed 10.12.2021).

Norberg-Schulz, C., Norberg-Schulz, A.M. (1992). *Genius loci: Landscape, environment, architecture, architectural documents*. Milan: Electa.

Pagnottella, P. (2002). L'Arsenale nel futuro di Venezia: Il progetto della Marina Militare, in Arsenale è/e Museo. *Quaderni Insula*, 11, pp. 33–38.

Pastor, V. (2017). *L'Arsenale di Venezia, progetti tentativo*. Padova: il Poligrafo.

Peleton, T., Gaggiotti, H., Case, P. (2018). *Origins of organizing*. Cheltenham: Elgar.

Perdomi, E. (2008). *L'Arsenale di Venezia – progetto di uno spazio espositivo di archeologia navale*. Tesi di laurea magistrale, IUAV, Venezia.

Pfeffer, J. (2009). Renaissance and renewal in management studies: Relevance regained, *European Management Review*, 6, pp. 141–148.

Polanyi, K. (1977). *The livelihood of man*. New York: Academic Press, Inc.

Rapp, R.T. (1976). *Industry and economic decline in seventeenth century Venice*. Cambridge, MA: Harvard University Press.

Romano, R. (1954). Aspetti economici degli armamenti navali veneziani. *Rivista Storica Italiana*, LXVI, pp. 39–67.

Samiolo, R. (2012). Commensuration and styles of reasoning: Venice, cost -benefit, and the defence of place. *Accounting, Organizations and Society*, 37, pp. 382–402.

Tarr, D.K. (ed) (2016). *Remaking post-industrial cities*. New York and London: Routledge.

Throsby, D. (2005). *Economia e cultura*. Bologna: Mulino.

Tonini, C., Lucchi, P. (a cura di) (2001). *Navigare e descrivere: Isolari e portolani del Museo Correr di Venezia (XV–XVIII secolo)*. Venezia: Marsilio.

UNESCO (2016a). *World heritage property Venice and its lagoon – (Italy) (C 394)*. http://docplayer.it/59752061-Format-for-the-submission-of-state-of-conserva tion-reports-by-the-states-parties-in-compliance-with-paragraph-169-of-the-operational-guidelines.html (Accessed 02.01.2018).

UNESCO (2016b). *Report of the joint Unesco/Icomos/Ramsar reactive monitoring mission to Venice and its lagoon.* http://whc.unesco.org/en/list/394/documents/ (Accessed 03.01.2018).

Zambon, S., Zan, L. (2007). Controlling expenditure, or the slow emergence of costing at the Venice Arsenal (1586–1633). *Accounting, Business & Financial History,* 17(1), pp. 105–128.

Zan, L. (2004a). Accounting and management discourse in protoindustrial settings: The Venice Arsenal in the turn of the XVI Century. *Accounting and Business Research,* 32(2), pp. 145–175.

Zan, L. (2004b). Toward a history of accounting histories. *European Accounting Review,* 3, pp. 255–310.

Zan, L. (2005). Future directions from the past: Management and accounting discourse in historical perspective. *Advances in Strategic Management,* 22, pp. 457–489.

Zan, L. (a cura di) (2018). *Per un futuro dell'Arsenale di Venezia.* Venezia: Cafoscarina.

Zan, L. (a cura di) (2019a). *Arsenale di Venezia: quale museo e quale accessibilità.* Venezia: Cafoscarina.

Zan, L. (2019b). History of management and stratigraphy of organizing: The Venice arsenal between tangible and intangible heritage. *Heritage.* www.mdpi.com/journal/heritage/special_issues/bah.

Zan, L. (2021). Il discorso del maneggio tra conoscenza, routine e impatti (sec VVI–XVII). Per una valorizzazione del significato intangibile dell'Arsenale. *Ateneo Veneto,* CCVIII(20/I), pp. 23–71.

Zan, L., Rossi, F., Zambon, S. (2006). *Il "discorso del maneggio": Pratiche gestionali e contabili all'Arsenale di Venezia, 1580–1643.* Bologna: il Mulino.

Zanetti, M. (2002). Un'idea per l'Arsenale. *Quaderni Insula,* pp. 5–6.

Index

Note: Page numbers in *italics* indicate a figure on the corresponding page.

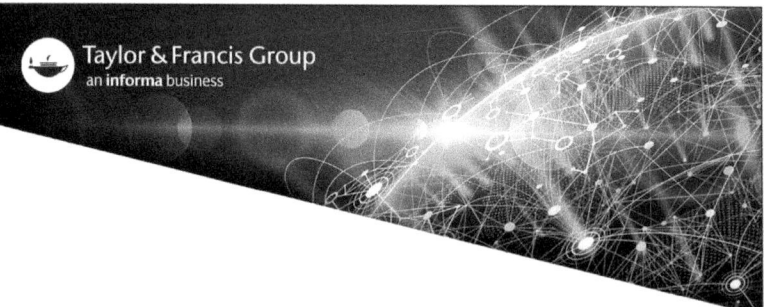